**Introduction to the
construction of class fields**

Introduction to the construction of class fields

HARVEY COHN

City University of New York

The right of the
University of Cambridge
to print and sell
all manner of books
was granted by
Henry VIII in 1534.
The University has printed
and published continuously
since 1584.

CAMBRIDGE UNIVERSITY PRESS

Cambridge

London   New York   New Rochelle

Melbourne   Sydney

Published by the Press Syndicate of the University of Cambridge
The Pitt Building, Trumpington Street, Cambridge CB2 1RP
32 East 57th Street, New York, NY 10022, USA
10 Stamford Road, Oakleigh, Melbourne 3166, Australia

First published 1985

Printed in the United States of America

Library of Congress Cataloging in Publication Data

Cohn, Harvey.

Introduction to the construction of class fields.

(Cambridge studies in advanced mathematics; 6)

Bibliography: p.

Includes index.

1. Class field theory.   I. Title   II. Series.
QA247.C634   1985   512′.32   84–27489
ISBN 0 521 24762 4

Contents

Preface

It is an elementary observation that an integral right triangle has an even area. Suppose the hypotenuse is prime.

Q. How do we determine from the prime value of the hypotenuse when the area is divisible by 4, 8, 16, or any higher power of 2?

A. We use class fields constructed by means of transcendental functions, of course!

The question might have been asked by Pythagoras in about 500 B.C. (but apparently was not) and the answer might have been given by Weber 2400 years later (but apparently was not). This text concentrates on the question and answer as a motivation for constructing class fields. The process of construction has gained interest lately partly because of the improved computer technology available. Yet the theory has benefited from this renewed interest in computation as much as the practical techniques (if not more so).

In meeting the special constructional objectives, this text provides a broad introduction to classical class field theory and modular function theory. This includes a more elementary review of algebraic number theory and Riemann surface theory for the purpose of emphasizing the relevant features and providing self-contained references.

The material consists of a year course for students already familiar with algebraic number theory and a semester course for students familiar with class field theory.

The first chapter sets up the motivation in terms of the "splitting" of primes, and under this motivation algebraic number theory is reviewed in Chapters 2, 3, and 4. In Chapter 5 the analytic concepts involving Riemann surfaces and modular equivalence are reviewed, particularly for genera 0 and 1. Chapters 6, 7, and 8 provide the transition from algebraic numbers to class fields (through the genus field concept). In Chapter 8 a simplified version of ideal transference and its relation to Galois groups is given (without introducing cohomology). Up to Chapter 8, the main proofs are referred to the references or sketched in some cases. From this point on, however, where the reader familiar with class field theory would presumably begin to assess new material, the proofs are relatively detailed.

The remaining chapters are more specialized. In Chapter 9 radicals are used for illustrations of unramified and ramified towers, and Hilbert's theory of units is introduced for the purpose. As an addendum the more recent concept of "governing field" is discussed. In Chapters 10 and 11 modular functions are used for singular moduli and for the class equation. This part is completely self-contained. The connection with Klein's icosahedron theory is indicated as a means of making the historical point that the construction of class fields is essential to many mathematical viewpoints. The concluding remarks are in the form of an epilogue rather than a survey of current theory. It is the author's hope that the reader will have a better grasp of classical results and will thereby be aided in seeing the present and creating the future.

The author's personal debts are always too personal to mention, but his professional debts are mainly to the students who had the patience to read and comment on badly hand-written and poorly copied early versions of the manuscript. Accordingly, thanks are due to Nicomedes Alonso, Mahmood Haghighi, Howard Pzena, Myong Shenefelt, and Clara Wajngurt. Finally, the author has benefited from a careful and critical reading of the manuscript by Jesse Deutsch.

H. C.

1

Preview: Quadratic forms and modular functions

The construction of class fields, particularly the use of transcendental methods for an inherently algebraic problem, is still challenging today. We illustrate this idea by following a classical problem born in the theorem of Pythagoras but recurrent over the centuries. For ease of discussion, we indulge in an oversimplification of ideas, which will be rendered more precise later on.

For algebraic number theory the fountainhead is a theorem of Fermat concerning a rational prime p (> 0) that is to be represented as a "Pythagorean hypotenuse," that is as the quadratic form in the integers x and y,

$$(1.1) \qquad p = x^2 + y^2.$$

Fermat's theorem states that this is possible if and only if $p = 2$ or p is of the form $4g + 1$ (for g a rational integer). The proof will be available later, but now we note merely that the theorem is derivable from the following statements about linear forms:

1. An even square is of the form $4g$.
2. An odd square is of the form $4g + 1$ (indeed $8g + 1$).

We can rewrite equation (1.1) more concisely (ignoring $p = 2$) by using a congruential notation and noting that x or y (say y) must be even. Then Fermat's theorem states:

$$(1.2) \qquad p = x^2 + 4y^2 \Leftrightarrow p \equiv 1 \bmod 4.$$

We might still go on to ask whether y [in equation (1.2)] is even. Actually, using the preceding statements 1 and 2, we can deduce from equation (1.2):

$$(1.3) \qquad p = x^2 + 16y^2 \Leftrightarrow p \equiv 1 \bmod 8.$$

[Indeed if y is odd in equation (1.2), then $p \equiv 5 \bmod 8$.] If we attempt to ask again when y is even in equation (1.3), we find (with Dirichlet) that the condition is no longer linear (i.e., congruential) on p:

$$(1.4) \quad p = x^2 + 64y^2 \Leftrightarrow p \equiv 1 \bmod 8 \quad \text{and} \quad w^4 \equiv 2 \bmod p.$$

(The last statement is that an integral w exists that satisfies the congruence shown; that is, a biquadratic root of 2 exists modulo p.)

A major result of class field theory (see Chapter 7) will be to reassure us that in equation (1.4) we are dealing with linear congruences not on p but on *divisors of p*. For now we succumb to the format of using conditions modulo p, rather than conditions on p, in the interest of uniformity. For instance, by a form of "quadratic reciprocity" (see Chapter 2),

$$u^2 + 1 \equiv 0 \bmod p \Leftrightarrow p \equiv 1 \bmod 4$$
$$v^2 - 2 \equiv 0 \bmod p \Leftrightarrow p \equiv \pm 1 \bmod 8.$$

(As before, the implication is that the written variables u and v have integral values.) Then the conditions in equations (1.2), (1.3), and (1.4) can be written as follows:

$$(1.5a) \qquad p = x^2 + 4y^2 \ \Leftrightarrow u^2 + 1 \equiv 0 \bmod p$$

$$(1.5b) \qquad p = x^2 + 16y^2 \Leftrightarrow \begin{cases} u^2 + 1 \equiv 0 \bmod p \\ v^2 - 2 \equiv 0 \bmod p \end{cases}$$

$$(1.5c) \qquad p = x^2 + 64y^2 \Leftrightarrow \begin{cases} u^2 + 1 \equiv 0 \bmod p \\ v^2 - 2 \equiv 0 \bmod p \\ w^2 - v \equiv 0 \bmod p. \end{cases}$$

The situation involving the congruential equations is later described as "normal" (meaning that the choice of roots of the congruence is indifferent). For instance, to illustrate equation (1.5c) with $p = 73 = 3^2 + 64$, we see $u \equiv \pm 27$ and 73. There are several choices of v and w, all equally valid; that is, $v \equiv 41$, $w \equiv \pm 25$ and $v \equiv -41$, $w \equiv \pm 18$. [The condition $w^4 \equiv 2$ in equation (1.4) appears as two steps in equation (1.5c) only to emphasize the inclusion of equation (1.5b).] Still, these results lack any apparent pattern.

Before showing that there is a pattern to these results, we introduce an algebraic concept, "splitting," which will be redefined with more precision in Chapter 3. We say that a prime "splits" in a finite extension field k over the rational numbers $\mathbf{Q}$ when every monic polynomial that defines an element of k (i.e., an algebraic integer of k) will split into linear factors modulo p. (We must exclude elements of k whose discriminants are divisible by p, but this is not important for now.) A more important fact (see Chapter 2) is that the splitting property must be proved for only one special primitive polynomial (whose root generates the field). Thus, equation (1.5a) is a condition for p to split $\mathbf{Q}(i)$. The splitting of a polynomial means in some crude way that each root "exists" or "is represented by an integer" modulo p [i.e., by u, v, w in equations (1.5a, b, c)]. All this must be made precise later on.

Thus we rewrite equations (1.5a, b, c) as follows:

(1.6a) $p = x^2 + 4y^2 \Leftrightarrow p$ splits in $\mathbf{Q}(i) (= k_2)$

(1.6b) $p = x^2 + 16y^2 \Leftrightarrow p$ splits in $\mathbf{Q}(i, \sqrt{2}) (= k_4)$

(1.6c) $p = x^2 + 64y^2 \Leftrightarrow p$ splits in $\mathbf{Q}(i, \sqrt[4]{2}) (= k_8)$.

The lack of an obvious pattern is seen in the next step:

(1.6d) $p = x^2 + 246y^2 \Leftrightarrow p$ splits in $\mathbf{Q}(i, \sqrt[8]{2} \cdot \sqrt{(1 + \sqrt{2})}) (= k_{16})$.

The fields k_n have their degrees shown as subscripts. They are called "ring class fields" for the quadratic forms on the left. (Reference is made to the "ring" $\mathbf{Z}[fi]$ whose general element $x + fiy$ is multiplied by its conjugate for $f = 2, 4, \ldots$, whereas the reference to "class" is deferred until later.) All this would be a futile exercise in nomenclature if not for a powerful result of Weber: **All ring class fields can be constructed explicitly by adjoining to the splitting field the values of certain transcendental functions ("modular functions") found by setting the argument equal to the root of the form.**

We recall that the roots of a form ρ and ρ' result from the factorization in $\mathbf{C}$ generally; that is,

$$ax^2 + bxy + cy^2 = a(x - \rho y)(x - \rho' y)$$

and the splitting field of the form is $\mathbf{Q}(\rho) [= \mathbf{Q}(\rho')]$.

The particular transcendental function involved is called the "modular invariant" $j(\tau)$, where τ is evaluated at ρ, the root of the quadratic form in question. The definition and evaluation of $j(\rho)$ are a major enterprise of the text and little can be said now beyond the formal expression (which is one of many; see Chapter 11):

(1.7a) $$j(\tau) = [1 + 256h(\tau)]^3/h(\tau)$$

(1.7b) $$h(\tau) = e(\tau) \prod_{m=1}^{\infty} [1 + e(m\tau)]^{24}$$

and $e(\tau)$ is the exponential function,

$$(1.7c) \qquad e(\tau) = \exp 2\pi i \tau \quad (= e^{2\pi i \tau}).$$

The function $e(\tau)$ has the familiar property that it is a function invariant under the transformation $\tau \to \tau + 1$. It is also algebraic for τ rational, taking on as values the roots of unity. The function $j(\tau)$ is invariant under the additional transformation $\tau \to -1/\tau$, and it takes on algebraic values for complex quadratic surd arguments τ [which have positive imaginary parts for convergence in equation (1.7b)].

We are hereby asserting that the previous results are all contained in the following statement:

$$(1.8) \qquad p = x^2 + 4.4^t y^2 \Leftrightarrow p \text{ splits in } \mathbf{Q}(i, j(2^{t+1}i)) \quad (= k_{2.2t}).$$

This prescription works for any principal form (i.e., a form that represents 1) as long as the discriminant d is negative. We write

$$(1.9a) \qquad F_d(x,y) = \begin{cases} x^2 - dy^2/4, & d \equiv 0 \bmod 4 \\ x^2 + xy - (d-1)y^2/4, & d \equiv 1 \bmod 4. \end{cases}$$

Then Weber's theorem states that for $d < 0$,

$$(1.9b) \qquad p = F_d(x, y) \Leftrightarrow p \text{ splits in } \mathbf{Q}(\rho, j(\rho)) \quad (= \text{RCF}\{d\}).$$

Here RCF$\{d\}$ denotes the ring class field for discriminant d and

$$(1.9c) \qquad \rho = \begin{cases} \sqrt{(d/4)} & d \equiv 0 \bmod 4 \\ (-1 + \sqrt{d})/2, & d \equiv 1 \bmod 4. \end{cases}$$

The annoyance of having two congruence cases for d can be somewhat alleviated by using the law $j(\tau) = j(\tau + 1)$ to write

$$(1.9d) \qquad \text{RCF}\{d\} = \mathbf{Q}(\sqrt{d}, j((d + \sqrt{d})/2)).$$

The field RCF$\{d\}$ is always abelian over $\mathbf{Q}(\sqrt{d})$, but it is not the most general such field. It is an important problem to construct the most general field possible by using transcendental functions (like j) evaluated for algebraic values (like ρ). In its most extended form this is "Hilbert's twelfth problem." Yet even Weber's comparatively narrow result is a quintessential theorem of the nineteenth century. Weber's theorem connects number theory with Galois theory as well as Riemann surface theory and algebraic geometry (which had all stood relatively aloof from it until then). Indeed, if we modify the problem to ask to what power $5^t \mid y$ (see Chapter 11), we can reconstruct Klein's approach to the icosahedron!

We conclude with a special result derived from Weber's theorem illustrating once more how rational number theory mysteriously requires transcendental methods. We use the additional notation $x \equiv a/b \bmod p$ to denote the solution to $ax \equiv b \bmod p$. Then the resulting equation (1.8) can be rewritten for $p \equiv 1 \bmod 4$ as follows:

(1.10)

$$p = x^2 + 4.4^t y^2 \Leftrightarrow \begin{cases} \text{a sequence of integers } r_1, \ldots, r_t \text{ is} \\ \text{definable modulo } p \text{ under the recursion} \\ r_1^2 \equiv 9/8 \\ r_{s+1}^2 \equiv (r_s + 3)^2/[8(r_s + 1)]. \end{cases}$$

This last result is necessarily related to the algebraic relation between $j(\rho)$ and $j(2\rho)$ developed in Chapter 10 (the modular equation). For now, we can check this result for small t against (say) equations (1.6a, b, c). For $t = 1$, we require that $\sqrt{9/8}$ (hence $\sqrt{2}$) exists modulo p. For $t = 2$, we just need $8(r_1 + 1)$ to be a perfect square. This leads to the assertion that

$$8(\sqrt{9/8} + 1) = \sqrt{8}(1 + \sqrt{2})^2$$

is also a perfect square [which leads to $\sqrt[4]{2}$ in equation (1.6c)].

Exercise

1.1. For the representation $p = x^2 + 2.4^t y^2$ the algorithm corresponding to equation (1.10) requires that $p \equiv 1$ or $3 \bmod 8$ and that we start with $r_1^2 \equiv 2$. Derive analogs to equations (1.6a, b, c). Consider the simultaneous representations (for small t) of

$$p = x^2 + 2.4^t y^2 = X^2 + 4.4^t Y^2.$$

Hint: For $t \leq 6$, the splitting field for p will have only the radicals $\sqrt{-1}$ and $\sqrt{2}, \sqrt[4]{2}, \ldots, \sqrt{1 + \sqrt{2}}, \sqrt[4]{1 + \sqrt{2}}, \ldots$.

Bibliographic note

The motivating illustration is found in Cohn (1981b); a proof of Weber's theorem is found in Hasse (1927b).

2

Early versions of class field theory

In this chapter we formalize the background for the curiosities of the preceding chapter, at least in terms of the perceptions that antedate the modern algebraic structures and Weber's theorem. This involves an informal survey of Kummer's approach to the concept of the splitting of primes and Gauss's and Dirichlet's approach to the genera of quadratic forms. We do not attempt to reconstruct the machinery to provide major proofs, however, since any such efforts will be amply preempted in the climactic theorems of Chapter 7. We do, however, provide a motivated review of elementary number theory and an indication of the importance of the discriminant, integrality, and related concepts in the later formulation.

It may be surprising at first that the concepts are harder to formulate in the early, elementary terms than they are in the more sophisticated language that evolved. Furthermore, the theorems are less natural in the "source language." For example, they are plagued with both special restrictions and exceptional cases, which later seem unnecessary. Yet it is this difficulty that justifies the escalation of the degree of abstraction in succeeding chapters.

For now, we find that elementary number theory has a consistent motivation built around the concepts. We see a coherent role for residuacity and reciprocity, the arithmetic progressions of primes, residue class groups, character theory, and so forth.

2.1. Rational definition of class field

Let $f(x)$ be a polynomial of degree m in $\mathbf{Z}[x]$:

$$f(x) = a_0 x^m + a_1 x^{m-1} + \cdots + a_{m'} \qquad (m \geq 1, a_0 a_m \neq 0).$$

If $f(x)$ is irreducible over **Z** (or **Q**), it is the unique "defining polynomial" of any of its roots, after we agree to eliminate common divisors [i.e., $(a_0, \ldots, a_m) = 1$] and set $a_0 > 0$. Such roots are "algebraic numbers."

2.1.1. Definition. *If $f(x)$ is the defining polynomial of its root θ, we call a_0 the "denominator" of θ. If $a_0 = 1$, or if $f(x)$ is monic, we call θ an algebraic "integer." In any case, $\psi = a_0\theta$ is an algebraic integer that satisfies*

$$\psi^m + a_0 a_1 \psi^{m-1} + \cdots + a_0^m a_m = 0.$$

A defining polynomial of θ is also called a generating or defining polynomial of the field $\mathbf{Q}(\theta)$. (It can always be chosen monic.)

2.1.2. Definition. *The prime p "splits" the polynomial $f(x)$ when $p \nmid a_0$ and there are m different rational (integral) roots $x_1, x_2, \ldots, x_m$ to the equation*

$$f(x) \equiv 0 \ mod \ p.$$

We say that each such root x_t "represents" a value of θ_t mod p for θ_t a conjugate of θ. We then say that p "splits the field" $K = \mathbf{Q}(\theta)$ for the generator θ.

Easily, $p \geq m$, but we can go further.

2.1.2a. Theorem. *The prime p splits the polynomial $f(x)$ exactly when $f(x)$ divides $x^p - x$ with coefficients taken modulo p.*

This is a consequence of Fermat's theorem, written as follows:

$$(2.1.2b) \qquad x^p - x \equiv \prod_{a=0}^{p-1} (x - a) \ mod \ p.$$

The equation (1.5a) now tells us that the set of primes of the form $x^2 + 4y^2$ splits the polynomial $x^2 + 1$. The same could be said about the form $x^2 + y^2$ if we exclude $p = 2$. In general, most "naturally" defined splitting sets will lead to exceptions, and the theory must allow the exceptions if they are finite in number.

2.1.3. Definitions for a set of primes P. *The set P is said to "split the polynomial" $f(x)$ when all $p \in P$ split $f(x)$, with finite exceptions (abbre-*

viated wfe). Thus, a finite set of primes splits any polynomial. Two sets of primes P_1 and P_2 that differ by a finite set of primes are called "equivalent" and are denoted by

$$P_1 = P_2 \quad wfe.$$

Analogously, we define an inclusion to hold with finite exceptions; that is, $P_1 \supseteq P_2$ wfe. The set P of all p that split f(x) is called "maximal" for f(x) when it contains all primes with the said property (wfe).

2.1.4. Definition. *A set of primes P "splits a field" K over $\mathbf{Q}$ when the defining polynomial for some generating θ in K is split by the set P. (We say, trivially, that any set P splits $\mathbf{Q}$.) Under this definition, each θ has a different set of exceptional p in P for which the polynomial defining θ might not split. The union of all primes that split a normal field for at least one generator of K is called "maximal" for $K/\mathbf{Q}$ (as is any equivalent set wfe). We denote this maximal set by*

(2.1.4a) $$P = Spl(K).$$

2.1.4b. Remark. Here we see one major difficulty of the elementary approach. The exceptional set of primes that split a field for one generator and not another can include any prime. For instance, if p splits K for the generator θ, it surely will not for the generator $p\theta$. In general, moreover, there is no best generator, so a unique maximal set cannot be defined conveniently at present. This can be done in Section 3.3 (using ideal theory).

2.1.5. Theorem. *Any set P that splits a normal field $K = \mathbf{Q}(\theta)$ for one generator will split it for any other generator. Indeed, for any element ϕ of K, generator or not, P will split the subfield $\mathbf{Q}(\phi)$ for ϕ. (The subfield need not be normal.)*

2.1.6. Remark. In preparing to prove the result in Theorem 2.1.5, we note that the definition of normality of K requires that if θ is a generator of K that has conjugates θ_i $(i = 1, \ldots, m)$, then

$$K = \mathbf{Q}(\theta_1) = \mathbf{Q}(\theta_2) = \cdots = \mathbf{Q}(\theta_m) \quad (\theta = \theta_1)$$

so for each i and j, $\theta_i \in \mathbf{Q}(\theta_j)$, and

(2.1.6a) $$\theta_j = v_j(\theta_1) = g_j(\theta_1)/M \quad (j = 1, \ldots, m)$$

for $v_j(x) \in \mathbf{Q}[x]$ and $g_i(x) \in \mathbf{Z}[x]$ with $M \in \mathbf{Z}^+$. [Note that M is the lowest common multiple (lcm) of various denominators. Thus, $v_1(x) = x$ but $g_1(x) = Mx$.] We do not yet assume that K is abelian. This would involve the further assumption of commutativity; that is, $v_i(v_j(\theta_1)) = v_j(v_i(\theta_1))$.

2.1.7. Lemma. *For a given generating polynomial $f(x)$ of a normal field K and for all p wfe, if x_1 is a root of $f(x) \equiv 0 \bmod p$, then the other roots are given by*

$$(2.1.8) \qquad x_t \equiv g_t(x_1)/M \bmod p \qquad (t = 1, \ldots, m).$$

In short, the representatives modulo p are related in the same manner as the roots when splitting occurs. Therefore, if $f(x)$ has one root modulo p, then it has m roots and p splits $f(x)$ (wfe, of course).

Proof (of Lemma 2.1.7). If a_0 is the denominator of the roots θ_i, then $a_0\theta_i$ are albegraic integers whose root discriminant (defined more formally in Chapter 4) is the rational integer

$$(2.1.9) \qquad d = \prod_{i>j} (a_0\theta_i - a_0\theta_j)^2.$$

We assert that the exceptional primes are merely some divisors of

$$(2.1.9a) \qquad D_0(\theta) = Ma_0 d.$$

For convenience, we assume that θ is replaced by the algebraic integer $a_0\theta$ as a generator of K (so we have $a_0 = 1$). We need some properties of algebraic integers (treated in Chapter 3). For instance, algebraic integers generate a ring; an algebraic integer of K lies in $\mathbf{Z}$ if and only if it is rational.

By elementary factorization, $Mx_t - g_t(x_1) = pw$ for some $w \in \mathbf{Z}$; hence, for each $t = 1, \ldots, m$,

$$(2.1.9b) \qquad \begin{aligned} x_t - \theta_t &= [g_t(x_1) - g_t(\theta_1)]/M + pw/M \\ &= [(x_1 - \theta_1)v + pw]/M \end{aligned}$$

where $v = G(\theta_1)$ for some $G(x) \in \mathbf{Z}[x]$. Here, v and w both depend on the value of t. (This result follows from the divisibility of $x^u - \theta^u$ by $x - \theta$.) Now $f(x) = N(x - \theta) = N(x - \theta_t)$. (Here the norm denotes the product over conjugates of θ; formal definitions are reserved for Chapter 3.) Thus,

$$M^m f(x_t) = N(M(x_t - \theta)) = N((x_1 - \theta)v + pw) \equiv f(x_1)Nv \bmod p.$$

Then easily, $f(x_i) \equiv 0$ implies $f(x_t) \equiv 0 \bmod p$. It remains to show that $x_i \not\equiv x_j \bmod p$ if $i \neq j$. We proceed to set

$$d = \prod_{i>j} (\theta_i - \theta_j)^2$$
$$= \prod_{i>j} (x_i - x_j + (\theta_i - x_i) - (\theta_j - x_j))^2.$$

By equation (2.1.9b),

$$d = \Pi \{x_i - x_j + [(x_1 - \theta_1)v + pw]/M\}^2.$$

We multiply out and expand the right-hand side to obtain

$$M^{m(m-1)}d \equiv M^{m(m-1)} \Pi (x_i - x_j)^2 + T \bmod p.$$

Here T is a rational integer. It is divisible by $(x_1 - \theta_1)$ and hence divisible by p. To see this last point, write

$$T = (x_1 - \theta_1)U(\theta_1)$$

for U an algebraic integer. Taking norms, we find

$$T^m = f(x_1)N(U)$$

and $f(x_1) \equiv 0 \bmod p$.   □

Proof (of theorem 2.1.5). Let P be a set of primes (not necessarily maximal) that splits the normal field $K = \mathbf{Q}(\theta)$ and let ϕ be another element of K. We show that P splits the field generated by ϕ (which would be K if ϕ is another generator). Then ϕ is one of n conjugates $\phi_1, \ldots, \phi_n$ that satisfies

$$\phi_t = h_t(\theta_1)/M \qquad (t = 1, \ldots, n)$$

for $h_t \in \mathbf{Z}[x]$ and $M \in \mathbf{Z}^+$. [For now all we need to know is that ϕ has a defining equation $s(\phi) = 0$ of degree n, where $n \mid m$.] Thus,

$$s(h_t(\theta_1)/M) = 0.$$

From the earlier discussion, $y_t \equiv h_t(x_1)/M$ provides the roots of $s(y) \equiv 0 \bmod p$. These roots are different by use of the discriminant $b_0^{n(n-1)}$ $\Pi_{i>j}(\phi_i - \phi_j)^2$ for b_0, the denominator of ϕ.   □

(Note that these proofs use the notation $x \equiv a/b \bmod p$, which means $bx \equiv a \bmod p$, as in Chapter 1.)

2.1.10. Theorem. *The maximal set of primes that splits a normal field is infinite.*

2.1.10a. Lemma. *For any nonconstant polynomial $f(x)$, the set of primes that divides the values of $f(x)$ and x varies in $\mathbf{Z}$ is infinite.*

Proof (of Lemma 2.1.10a). Otherwise let a finite set of primes $p_1, \ldots, p_r$ (of product R) be the only primes that divide values of $f(x)$. Let $f(a) \neq 0$ and consider the values of $x = a + tRf(a)$ for t variable in $\mathbf{Z}$. For these values, $f(x) \equiv f(a) \bmod Rf(a)$, so p_i (as a divisor of R) can divide $f(x)$ only to the same power that it divides $f(a)$. Hence a contradiction occurs with $|f(x)| = |f(a)|$ as t (and x) $\to \infty$. $\square$

Theorem 2.1.10 now follows from the observation that if the roots of $f(x)$ generate a normal field and (wfe) if p is a prime for which $f(x) \equiv 0 \bmod p$ is solvable, then p splits $f(x)$.

2.1.10b. Remark. Theorem 2.1.10 would also hold for a nonnormal field (because it can always be imbedded in its normal closure).

2.1.11. Theorem. *The maximal set that splits a normal field is equivalent to the maximal set that splits it for any generator.*

For the proof, note that these sets differ at most in the primes that divide $D_0(\theta)$ [in equation (2.1.9a)].

2.1.12. Lemma. *If $K = K_1 \times K_2$ and $K_1 \cap K_2 = \mathbf{Q}$ (for K_1 and K_2 normal), then P splits K if and only if it splits each K_i.*

Proof. Let $K_1 = \mathbf{Q}(\theta)$ and $K_2 = \mathbf{Q}(\phi)$, where θ is defined by $f(\theta) = 0$ of degree $m = [K_1 : \mathbf{Q}]$ and ϕ is defined by $s(\phi) = 0$ of degree $n = [K_2 : \mathbf{Q}]$. Then $nm = [K : \mathbf{Q}]$, so let ω generate $K/\mathbf{Q}$ and satisfy an equation of degree nm. We can write $\omega = F(\theta, \phi)/M'$ for $F(x, y) \in \mathbf{Z}[x, y]$ and $M' \in \mathbf{Z}^+$. As before, with a suitable common denominator M,

$$(2.1.12a) \quad \theta_t = g_t(\theta_1)/M, \quad t = 1, \ldots, m \quad \text{and} \quad g_t(x) \in \mathbf{Z}[x]$$
$$(2.1.12b) \quad \phi_u = h_t(\phi_1)/M, \quad u = 1, \ldots, n \quad \text{and} \quad h_u(x) \in \mathbf{Z}[x].$$

Then if x_1 represents θ_1 and y_1 represents $\phi_1 \bmod p$, it is clear from the earlier discussions that

$$W_{tu} = F(g_t(x_1)/M, h_u(y_1)/M)$$

will represent mn different values of $\omega \bmod p$. They are different as a consequence of the earlier discriminant discussion; that is, $\Pi (\omega_{tu} - \omega_{t'u'})^2 \neq 0$ for a product over distinct pairs (t, u) and (t', u'). $\square$

2.1.13. Corollary. *Lemma 2.1.12 is true even if* $K_1 \cap K_2 \neq \mathbf{Q}$, *as long as both subfields are normal.*

Proof. The difference in proofs is that $f(\theta) = 0$ defines θ on $\mathbf{Q}$ but $s(\phi) = 0$ then defines ϕ over $K_1 = \mathbf{Q}(\theta)$. This defining equation for ϕ (and for K/K_1) will lie not in $\mathbf{Q}[x]$ but in $K_1[x]$. Nevertheless, θ_1 and ϕ_1 still represent integers mod p whenever K and K_1 split over $\mathbf{Q}$, and the proof proceeds as before with equation (2.1.12b) replaced by

$$(2.1.13a) \qquad \phi_u = h_u(\phi_1, \theta_1)/M. \qquad \square$$

2.1.14. Definition. *If, for a set of primes P, a maximal normal field K exists that is split by P and if P in turn is maximal (wfe) for $K/\mathbf{Q}$, then $K/\mathbf{Q}$ is called a "class field" for P. (The maximality of K means that K must include all normal fields split by P. This makes K unique if it exists.) We then write*

$$(2.1.15) \qquad K = CF\{P\}.$$

2.1.16. Lemma. *If $P_1 \supseteq P_2$ wfe, then $CF\{P_1\} \subseteq CF\{P_2\}$ (a contravariant relation), assuming the class fields exist. Likewise,*

$$(2.1.16a) \qquad CF\{Spl(K)\} \supseteq K.$$

2.1.17. Remark. Among the various "adverse" possibilities implied by Definition 2.1.14 are the following:
1. P can be "so large a set" that no field (or irreducible) polynomial of degree greater than one can be split by it. Then $CF\{P\}$ does not exist except in the trivial case where $P =$ (all primes wfe) and $CF\{P\} = \mathbf{Q}$.
2. P can be "so small a set" that there are (normal) fields of arbitrarily high degree split by P. Then $CF\{P\}$ does not exist (since we exclude fields of "infinite degree").
3. The maximal (normal) field that P splits is finite, but this field is split by infinitely many other primes, so $CF\{P\}$ does not exist. Counterbalancing these difficulties we have the following result, which is proved in Chapter 4 (Theorem 4.3.9).

2.1.18. Theorem. *If P is defined as the maximal set of primes that split a normal field K, then $CF\{P\} = K$; that is,*

$$CF\{Spl(K)\} = K.$$

For instance, the opening result of Fermat is that a certain set of primes may be described in two different ways:

(2.1.18a) $P = \{p = x^2 + y^2\} = \{p = 2 \text{ or } p \equiv 1 \bmod 4\}.$

This result is (nowadays) very elementary; so is the result that P splits $\mathbf{Q}(i)$. Therefore, if CF$\{P\}$ exists, it contains $\mathbf{Q}(i)$. The result we want goes further, namely, CF$\{P\}$ exists and (by Theorem 2.1.18)

(2.1.18b) $$\mathrm{CF}\{P\} = \mathbf{Q}(i).$$

2.1.19. Remark. As in equation (2.1.18a), some sets are representable by both linear congruences and quadratic forms. More will be said of this in the next section. The representations by quadratic forms in general lead to more interesting and more historically meaningful results. In the meantime, we remark that the existence of CF$\{P\}$ is not assured for every "reasonable sounding" set of primes P (even P represented by a linear congruence or quadratic form). The class fields will accordingly become the primary objects and the set of primes will become an incidental property as the theory unfolds.

2.2. Class fields for quadratic forms

The general theory of quadratic forms is best appreciated by the structures introduced in Chapter 8, but to illustrate the classical cases of class fields, we give an elementary survey using only the principal quadratic form, with variables and coefficients in $\mathbf{Z}$. It is defined, as in Chapter 1, by

(2.2.1) $F_d(x, y) = \begin{cases} x^2 - (d/4)y^2 & (d \equiv 0 \bmod 4) \\ x^2 + xy - [(d-1)/4]y^2 & (d \equiv 1 \bmod 4). \end{cases}$

Here d cannot be a perfect square, but it may have square factors. By completing the square and changing variables, we find (for $X, Y \in \mathbf{Z}$),

(2.2.2) $$4F_d(x, y) = X^2 - dY^2.$$

2.2.3. Definition. *A "fundamental" discriminant is one from which no perfect square can be removed to produce another discriminant (i.e., a value of $b^2 - 4ac$ for integral a, b, c). We write*

(2.2.4) $$d = d_f = f^2 d_1$$

where d_1 is fundamental and f is called the "order" of $F_d(x, y)$. Thus, d_1 has no odd or even square factors except possibly 4; furthermore, either d_1 is odd and congruent to 1 mod 4 or $d_1/4$ is congruent to -1 or 2 mod 4.

The form $F_d(x, y)$ can be factored in the field

$$(2.2.5) \qquad k = \mathbf{Q}(\sqrt{d}) = \mathbf{Q}(\sqrt{d_1}),$$

a field shared by all forms of the same fundamental discriminant d_1. (It is called the "splitting" field of the form.) Thus,

$$(2.2.6a) \qquad F_d(x, y) = (x - \rho y)(x - \rho' y)$$

$$(2.2.6b) \qquad \rho = \begin{cases} \sqrt{d}/2 \\ (-1 + \sqrt{d})/2 \end{cases} \qquad \rho' = \begin{cases} -\sqrt{d}/2 \\ (-1 - \sqrt{d})/2 \end{cases} \qquad d \equiv \begin{cases} 0 \bmod 4. \\ 1 \end{cases}$$

We now consider in succession four sets of primes P_i. [In each case we exclude the primes $p \mid 2d$ for convenience, although they are often a legitimate part of the theory. The variables continue to be rational integers subject to the conditions $(p, xy) = (z, 2d) = 1$.]

$$(2.2.7a) \qquad P_1 = \{p : zp = F_d(x, y) \text{ solvable}\}$$

$$(2.2.7b) \qquad P_2 = \{p : z^2p = F_d(x, y) \text{ solvable}\}$$

$$(2.2.7c) \qquad P_3 = \{p : p^{2t+1} = F_d(x, y) \text{ solvable}\}$$

$$(2.2.7d) \qquad P_4 = \{p : p = F_d(x, y) \text{ solvable}\}.$$

Clearly, $P_1 \supseteq P_2 \supseteq P_3 \supseteq P_4$. Thus, if (as we shall later know) class fields exist, then easily

$$(2.2.8) \qquad \mathrm{CF}\{P_1\} \subseteq \mathrm{CF}\{P_2\} \subseteq \mathrm{CF}\{P_3\} \subseteq \mathrm{CF}\{P_4\}.$$

For a few familiar cases with small d, all these fields are equal to k (see Remark 2.3.9), but for general d, they are different – and famous:

Field	Symbol	Designation
CF$\{P_1\}$	k	Splitting field of F_d (Chapter 2)
CF$\{P_2\}$	k_g	Genus (ring class) field of F_d (Chapters 2 and 8)
CF$\{P_3\}$	$\mathrm{RCF}_2\{d\}$	Two-Sylow ring class field of F_d (Chapter 9)
CF$\{P_4\}$	$\mathrm{RCF}\{d\}$	Ring class field of F_d (Chapters 1 and 8)

(If d is fundamental, the "ring class" fields are designated as "Hilbert" class fields.)

2.2.9. Definition. *The "Legendre-Kronecker quadratic residue" symbol is defined for p an odd prime (by Legendre) as follows:*

$$(2.2.9a) \quad (a/p) = \begin{cases} 1 \text{ if } x^2 \equiv a \bmod p \text{ is solvable for } x \in Z, \quad p \nmid a \\ -1 \text{ if } x^2 \not\equiv a \bmod p \text{ for } x \in Z, \quad p \nmid a \\ 0 \text{ if } p \mid a. \end{cases}$$

For d a discriminant (only) there is an extended (Kronecker) symbol:

$$(2.2.9b) \qquad (d/2) = \begin{cases} 1 \text{ if } d \equiv \bmod 8 \\ -1 \text{ if } d \equiv 5 \bmod 8 \\ 0 \text{ if } d \equiv 0 \bmod 4. \end{cases}$$

With m composite, $m = \pm \Pi\, p_i^{e_i}$, we define the "Jacobi symbol" multiplicatively (using either type, extended or not) as

$$(2.2.10) \qquad (a/m) = \Pi\, (a/p_i)^{e_i}.$$

(This ignores the sign of m, not a.) If m is even, this symbol requires that a be a discriminant. All symbols (a/b) are multiplicative in both the "numerator" and the "denominator"; indeed the discriminants are a multiplicative semigroup of Z.

2.2.11. Euler's lemma. *For p an odd prime,*

$$(a/p) \equiv a^{(p-1)/2} \bmod p.$$

2.2.12. Theorem. *Let* $P_1^* = \{p : (d/p) = 1\}$. *Then* $P_1 = P_1^*$ *wfe.*

Proof. Write $4pz = X^2 - dY^2$, excluding $p \mid 2d$. Thus $p \in P_1$ implies $d \equiv (X/Y)^2 \bmod p$. Conversely, if $d \equiv X^2 \bmod p$, then $4pz = X^2 - d$ as long as we choose X (or $X + p$ instead) to be odd when d is odd. This representation leads back to P_1^*. $\qquad\square$

2.2.13. Corollary. *The set* P_1^* *splits* $k = Q(\sqrt{d})$, *and indeed when d is fundamental it contains exactly all primes that do so with regard to the generator* ρ *[see equations (2.2.6a, b)]. In other words, it is maximal.*

2.2.14. Theorem. *The class field* $CF\{P_1\} = k$.

This is a deep result (see Theorem 2.1.18), but strangely enough, historically speaking, it attracted attention largely because of a more difficult result involving P_2. We start with $z^2p = F_d(x, y)$ for $p \nmid 2d$. Then we have, as in equation (2.2.2),

$$(2.2.15a) \quad 4z^2p = X^2 - dY^2, \qquad d \text{ odd or even } (z \text{ odd})$$
$$(2.2.15b) \quad z^2p = x^2 - (d/4)y^2, \qquad d \text{ even } (z \text{ odd}).$$

We consider the first relation for an odd prime q dividing d:

(2.2.16a) $4z^2p \equiv X^2 \bmod q, \quad q \mid d, \quad q \nmid z$

(2.2.17a) $(p/q) = 1.$

Relation (2.2.15b) can also be taken mod 8 (so $z^2 \equiv 1$) but it still breaks up into several cases if we start from

(2.2.16b) $p \equiv x^2 - (d/4)y^2 \bmod 8.$

Therefore, for d even,

(2.2.17b) $d/4 \equiv \begin{cases} 0 \bmod 8 \Rightarrow p \equiv 1 \bmod 8 \\ -1, 3, 4 \bmod 8 \Rightarrow p \equiv 1 \bmod 4 \ (\equiv 1, 5 \bmod 8) \\ -2 \bmod 8 \Rightarrow p \equiv 1, 3 \bmod 8 \\ 2 \bmod 8 \Rightarrow p \equiv 1, -1 \bmod 8 \\ 1, 5 \bmod 8 \Rightarrow \text{(no conclusion on } p). \end{cases}$

Thus we can define a new set of primes (of course, depending on d):

(2.2.18) $P_2^* = \{p : (p/q) = 1 \text{ for odd prime } q \mid d, \quad p \equiv a \bmod 8\}$

where a is selected for d even according to system (2.2.17b). Clearly, $P_2 \subseteq P_2^*$.

2.2.19. Genus theorem of Gauss and Dirichlet. *For a given discriminant d, P_2 and P_2^* are equivalent. The only exceptional primes are divisors of $2d$.)*

This result will be proved later as a result of class field theory (see Theorem 8.2.15). It was originally proved as an elementary result of the depth of quadratic reciprocity (see Exercise 2.3).

2.2.20. Quadratic reciprocity theorem (Gauss). *Let p and q be odd primes and let us denote $q^* = q(-1)^{(q-1)/2}$ so that*

$$q^* = \pm q \equiv 1 \bmod 4.$$

Thus, $3^* = -3, 5^* = 5,$ and so on. Then the main theorem is

(2.2.20a) $(p/q) = (q^*/p).$

The supplementary theorems are

(2.2.20b) $(-1/q) = (-1)^{(q-1)/2}$

(2.2.20c) $(2/q) = (-1)^{(q^2-1)/8}.$

Thus, $(p/q) = (q/p)$ unless both p and q are $\equiv -1$ mod 4, and then $(p/q) = -(q/p)$. Also, the residues mod 8 are characterized by

$$(2.2.21a) \quad \begin{cases} p \equiv 1 \text{ mod } 8 \leftrightarrow (-1/p) = (2/p) = 1 \\ p \equiv 1, 5 \text{ mod } 8 \leftrightarrow (-1/p) = 1 \\ p \equiv 1, 7 \text{ mod } 8 \leftrightarrow (2/p) = 1 \\ p \equiv 1, 3 \text{ mod } 8 \leftrightarrow (-2/p) = 1. \end{cases}$$

According to equation (2.2.17a), P_2^* splits $\mathbf{Q}(\sqrt{q^*})$ for each odd prime $q \mid d$ and also splits another K_0 of varying types,

$$(2.2.21b) \quad \begin{cases} d \text{ odd or } d/4 \equiv 1 \text{ mod } 4 \leftrightarrow K_0 = \mathbf{Q} \\ d/4 \equiv -1, 3, 4 \text{ mod } 8 \leftrightarrow K_0 = \mathbf{Q}(i) \\ d/4 \equiv -2 \text{ mod } 8 \leftrightarrow K_0 = \mathbf{Q}(\sqrt{-2}) \\ d/4 \equiv 2 \text{ mod } 8 \leftrightarrow K_0 = \mathbf{Q}\sqrt{2}) \\ d/4 \equiv 0 \text{ mod } 8 \leftrightarrow K_0 = \mathbf{Q}(i, \sqrt{2}). \end{cases}$$

The composite field in which P_2^* splits is now as follows:

$$(2.2.22) \qquad k_g = \mathbf{Q}(\sqrt{q_1^*}) \times \cdots \times \mathbf{Q}\sqrt{q_t^*}) \times K_0$$

where $q_1, \ldots, q_t$ are the odd-prime divisors of d.

2.2.23. Theorem (Gauss–Dirichlet). *The genus field $k_g = CF\{P_2\} = CF\{P_2^*\}$.*

2.3. Dirichlet characters

In many cases, primes in P are determined by arithmetic progressions. This is now formalized to give insight into the more general situation that will arise.

2.3.1. Theorem of Jacobi on quadratic reciprocity. *For the Jacobi symbol (b/a) with b and a odd and positive and $a^* = a(-1)^{(a-1)/2} \equiv 1$ mod 4,*

$$(2.3.1a) \qquad (b/a) = (a^*/b)$$
$$(2.3.1b) \qquad (-1/a) = (-1)^{(a-1)/2}$$
$$(2.3.1c) \qquad (2/a) = (-1)^{a^2-1)/8}.$$

If $a = a^* = d$ (a positive discriminant), then equation (2.3.1a) holds with b even. Thus, consistent with equation (2.2.20a),

$$(2.3.1d) \qquad (2/d) = (d/2), \qquad \text{odd } d > 0.$$

2.3.2. Definition. *A "character" mod m is a multiplicative function f(x)*
for x in $\mathbf{Z}^+$ *of period m in x with values (in C):*

(2.3.2a) $f(xy) = f(x)f(y)$
(2.3.2b) $f(x) = f(y)$, *if* $x \equiv y \bmod m$.

We require that f(x) not be identically vanishing [so f(1) = 1]. We call
m a "primitive" period if there is no smaller modulus (>1) for which
the periodicity holds. We call two characters, f(x) and g(x), "equivalent"
if they agree except for values of x divisible by a finite set of primes.
This makes for equivalence classes of characters. A "unit" character is
one where f(x) = 1 for all x or a character equivalent to it.

A character $f(x)$ cannot vanish for $x = w$ if $(w, m) = 1$. [Otherwise,
$f(x)$ vanishes identically, since if a u exists for which $uw \equiv 1 \bmod m$,
then $f(1) = f(u)f(w)$.] On the other hand, $f(m) = 0$ for a nonunit character, since $f(m) = f(mx) = f(m)f(x)$ for all x. If $(a, m) > 1$, the vanishing is governed by the following result:

2.3.3. Lemma. *For a nonunit character there is a primitive period m >*
1 that is the greatest common divisor (gcd) of all periods of equivalent
characters. There is an equivalent (so-called primitive) character of
period m such that f(a) = 0 exactly when (a, m) > 1.

2.3.4. Definition. *A Dirichlet character is the character*

$$\chi_d(x) = (d/x)$$

for d a discriminant, and the special case $\chi_*(x) = (m^2/x)$.

2.3.4a. Lemma. *The character* $\chi_d(x)$ *has primitive period* $d_1 = d/f^2$,
where d_1 *is the fundamental discriminant for d. The character* $\chi_*(x)$ *is,*
of course, equivalent to the unit character.

2.3.5. Theorem. *The only real characters are Dirichlet characters.*

2.3.6. Theorem. *The real characters can be extended to negative x by*
$\chi_*(-1) = 1$ *and* $\chi_d(-1) = $ sgn d *in a manner that preserves both peri-*
odicity and multiplicativity. In fact, these characters can be extended to
all rationals x/y where (d, xy) = 1 by multiplicativity for $\chi_d(\ldots)$ *and*
by setting $\chi_*(\ldots) = 1$.

These results are classical and are given as a guide for later purposes rather than for immediate use. They can be deduced from the structure of the multiplicative residues mod m and Jacobi reciprocity.

In the current context, we are dealing with sets of the type

$$(2.3.7) \qquad P[d] = \{p : (d/p) = 1\} = \{p : \chi_d(p) = 1\}.$$

Indeed, in Theorem 2.2.12 and equation (2.2.22) we saw that

$$(2.3.8a) \qquad P_1 = P[d] \quad \text{wfe}$$
$$(2.3.8b) \qquad P_2 = P[q_1^*] \cap \cdots \cap P[q_7^*] \cap P_0 \quad \text{wfe}$$

where

$$(2.3.8c) \qquad P_0 = \begin{cases} P[1], & d \text{ odd or } d/4 \equiv 1 \bmod 4 \\ P[-4], & d/4 \equiv -1, 3, 4 \bmod 8 \\ P[8], & d/4 = 2 \bmod 8 \\ P[-8], & d/4 \equiv -2 \bmod 8 \\ P[8] \cap P[-4], & d/4 \equiv 0 \bmod 8. \end{cases}$$

(Here $P[1]$ is a convenient symbol for the set of all primes.) The exceptions involved in wfe are only from $p \mid 2d$.

2.3.9. Remark. We can now see that the main burden of class field theory is to show how this situation generalizes. For instance, P_3 and P_4 cannot be expressed in terms of sets like $P[d]$ in general. In fact, this situation holds only when $CF\{P_3\}$ or $CF\{P_4\}$ is abelian over $\mathbf{Q}$ (as are $CF\{P_1\}$ and $CF\{P_2\}$; see Theorems 2.2.14 and 2.2.23). We learn that all these class fields are nevertheless abelian over k, so a theory of class fields over k (not $\mathbf{Q}$) and a theory of arithmetic progressions in integers of k (not $\mathbf{Z}$) must be developed. This will emerge in Chapter 7 after elaborate algebraic preparation.

2.4. Appendix on rational number theory

The residues modulo m (for $m \geq 1$) are an additive group denoted by $(\mathbf{Z}/m\mathbf{Z})^+$, or more conveniently by

$$(2.4.1) \qquad C(m) = \text{cyclic group of order } m.$$

The residues modulo m relatively prime to m are a multiplicative group $(\mathbf{Z}/m\mathbf{Z})^*$ of order $\phi(m)$ (Euler phi-function). This group has a cyclic decomposition governed by prime factors of m. Actually,

$$(2.4.2) \qquad (\mathbf{Z}/m_1 m_2 \mathbf{Z})^* = (\mathbf{Z}/m_1 \mathbf{Z})^* \times (\mathbf{Z}/m_2 \mathbf{Z})^*$$

if $(m_1, m_2) = 1$ (Chinese remainder theorem). Thus,

(2.4.3) $\phi(m_1 m_2) = \phi(m_1)\phi(m_2)$

and the residues modulo $m_1 m_2$ can be paired off uniquely as a Cartesian product

(2.4.4) $(x \bmod m_1 m_2) = (x \bmod m_1, x \bmod m_2)$.

For $m = p^e$, $\phi(m) = p^e - p^{e-1}$. Thus, in general,

(2.4.5) $\phi(m) = m \prod_{p|m} (1 - 1/p) = \sum_{d|m} m\mu(d)/d$

for primes p and (general) divisors d in m. Here $\mu(d)$ is the "Moebius μ-function" (nonvanishing only for square-free n)

(2.4.6) $\mu(n) = \begin{cases} 1 & n = 1 \\ 0 & n \text{ has square factor} > 1 \\ (-1)^t & n = p_1 \dots , p_t \text{ (distinct primes)}. \end{cases}$

For $m = p^e$ (p odd), $\phi(m) = p^{e-1}(p - 1)$ and

(2.4.7) $(\mathbf{Z}/p^e\mathbf{Z})^* = C(\phi(p^e))$.

There are $(\phi(p^e))$ generators of $C(\phi(p^e))$ modulo $\phi(p^e)$ called "primitive roots" g. They satisfy the property that for the $\phi(p^e)$ values of $x \bmod p^e$ where $(x, p) = 1$, the congruence

(2.4.8) $x \equiv g^t \bmod p^e, \qquad 0 \le t \le \phi(p^e) - 1$

defines a one-to-one relation between x and t.

For powers of two there is the special case:

(2.4.9) $p^e \begin{cases} = 2, & \phi(2) = 1, & (\mathbf{Z}/2\mathbf{Z})^* = C(1) \\ = 4, & \phi(4) = 2, & (\mathbf{Z}/4\mathbf{Z})^* = C(2) \\ \ge 8, & \phi(2^e) = 2^{e-1}, & (\mathbf{Z}/2^e\mathbf{Z})^* = C(2) \times C(2^{e-2}). \end{cases}$

Thus, we can write for all $\phi(2^e)$ odd values of $x \bmod 2^e$,

(2.4.10) $x \equiv (-1)^t 5^u, \qquad 0 \le t \le 1, \quad 0 \le u \le 2^{e-2} - 1$

defining a one-to-one relation between x and (t, u). The subgroup where $x \equiv 1 \bmod 4$ ($t = 0$) is, of course, cyclic.

The cyclic character of $(\mathbf{Z}/p\mathbf{Z})^*$ (odd p) and its even order lead to the result $(a/p)(b/p) = (ab/p)$ in Legendre symbols, since even powers of g in equation (2.4.8) are precisely the quadratic residues. Thus, we see that the real characters are precisely as described in Theorem 2.3.5 of Dirichlet. A real character $\chi(x)$ is defined mod m as the product of real characters mod p^e, where p^e are the exact prime powers that factor m (by the Chinese remainder theorem). For p odd, the real characters are obviously quadratic characters, since the characters are isomorphic with the group and they define the only subgroup of index 2. For $p = 2$ and $p^e \geq 8$ (for example) the real characters can be identified by their action on 5 and -1 as follows:

$\chi(5)$	$\chi(-1)$	$\chi(x)$	*Period*
1	1	$(1/x)$	1
1	-1	$(-4/x)$	4
-1	1	$(8/x)$	8
-1	-1	$(-8/x)$	8

The characters mod m form a group under multiplication that is dual (isomorphic) to $(\mathbf{Z}m\mathbf{Z})^*$. Thus, there are $\phi(m)$ characters of which only 2^{t+a} are real, where t is the number of odd primes that divide m and $a = 0$ if $4 \nmid m$, $a = 1$ if $4 \mid m$ but $8 \nmid m$, and $a = 2$ if $8 \mid m$.

Finally, the theorem on residues mod p, otherwise expressed, states that the elements of $(\mathbf{Z}/p\mathbf{Z}) = \mathbf{F}_p$ form a finite field of p elements (order p). Under multiplication, its nonzero elements $(\mathbf{Z}/p\mathbf{Z})^*$ form a group of order $(p - 1)$. This is the only field of order p.

A much deeper result is that the only finite fields of any order are fields of order p^s that are extensions of $\mathbf{F}_p$ of degree s formed by adjoining the root of an irreducible polynomial (of degree s) over $\mathbf{F}_p$,

$$(2.4.11) \qquad f(x) = x^s + a_1 x^{s-1} + \cdots + a_s = 0.$$

(The coeffficients a_i can, of course, be interpreted as integers modulo p or as elements of F_{p^r}.) Then if θ is a root of equation (2.4.11), $\mathbf{F}_p(\theta) = \mathbf{F}_q$, a field of $q = p^s$ elements uniquely determined, regardless of the choice of polynomial $f(x)$. Such polynomials always exist for any q and are identified as the polynomials that divide $x^q - x \bmod p$ but do not divide $x^{q/p} - x \bmod p$. We can therefore characterize q as the smallest power of p for which all elements of $\mathbf{F}_q$ satisfy $x^q = x$. Here we are using the "Frobenius" automorphism of $\mathbf{F}_q$:

$$(2.4.12) \qquad T : x \to x^p.$$

It preserves only $\mathbf{F}_p$ because $x^p - x$ has only p roots, but the iteration T^s will preserve $\mathbf{F}_q$ and indeed will be the first iteration to do so.

As a consequence of these considerations, the polynomial (2.4.11) irreducible modulo p is necessarily normal with Galois group cyclic of order s. The roots of $f(x)$ in $\mathbf{C}$ are represented by elements of $\mathbf{F}_q$ analogously with the splitting process we defined earlier when the irreducible factors are of degree $(s =) 1$. There the roots are represented by elements mod p and the group action (2.4.12) is trivial.

Exercises

2.1. Complete the proofs of Lemma 2.1.12 and Corollary 2.1.13 (particularly with regard to using the discriminant to show that W_{tu} takes mn different values).

2.2. Show that P_1 and P_1^* do not disagree when d is odd as far as $p = 2$ is concerned; that is, $2z = x^2 + xy - (d - 1)y^2/4$ is solvable when $d \equiv 1 \bmod 8$ only.

2.3. Deduce quadratic reciprocity from the genus theorem in the form 2.2.19. From equation (2.2.20b), if $p \equiv 1$, $q \equiv \pm 1 \bmod 4$ for $d = p$, $(q/p) = 1 \Rightarrow 4qz^2 = x^2 - py^2 \Rightarrow py^2 = x^2 - 4qz^2 \Rightarrow (p/q) = 1$. For $d = q^*$, $(p/q) = 1 \Rightarrow 4pz^2 = x^2 - q^*y^2 \Rightarrow q^*y^2 = x^2 - 4pz^2 \Rightarrow (q^*/p) = 1$. Finally, let $d = q_1q_2$, $q_1 \equiv q_2 \equiv -1 \bmod 4$. Then if $(q_1/q_2) = 1$, $4q_1z^2 = x^2 + q_2y^2 \Rightarrow -q_2y^2 = x^2 - 4q_1z^2 \Rightarrow (-q_2/q_1) = 1 \Rightarrow (q_2/q_1) = -1$.

2.4. (a) Verify the cases in equation (2.2.17b). (b) For the case where d is a fundamental discriminant, show that with $(d/p) = 1$, the even characters in equation (2.2.17b) can be deduced from the odd characters $(q_i^*/p) = 1$.

2.5. For the case where d is a fundamental discriminant, deduce the genus theorem $(P_2 = P_2^*$ wfe) from Legendre's theorem, which follows: If a, b, c are square-free integers not all of the same sign and $(a, b) = (b, c) = (c, a) = 1$, then the equation $ax^2 + by^2 + cz^2 = 0$ is solvable for nonzero integers if and only if it is so solvable mod q for each odd prime q that divides abc.

Hilbert introduced a "norm residue" symbol $(a, b/p)$ that is always defined for $a, b \in \mathbf{Z}$ $(ab \neq 0)$ and has values of ± 1 only:

$(a, b/p) = 1$ if a or b is a perfect square or $a \equiv F_d(x, y) \bmod p^t$

for F_d the principal form (i.e., norm) for the field $\mathbf{Q}(\sqrt{b})$ of discriminant d and all $t > 0$;

$$(a, b/p) = -1 \text{ otherwise.}$$

This symbol is evaluated by the following rules:

$(a, b/p) = 1$, p odd, $(ab, p) = 1$
$(a, p/p) = (p, a/p) = (a/p)$, p odd, $(ab, p) = 1$

$$(a, b/2) = (-1)^{(a-1)/2}(-1)^{(b-1)/2}$$
$$(a, 2/2) = (2, a/2) = (-1)^{(a^2-1)/8}$$

It has the further properties:

$$(-a, a/p) = 1$$
$$(a, b/p) = (b, a/p)$$
$$(aa', b/p) = (a, b/p)\,(a', b/p)$$
$$(a, bb'/p) = (a, b/p)\,(a, b'/p)$$

2.6. Verify that the genus symbols for the fundamental discriminant d are $\chi_i(m) = (m, d/p_i)$ where p_i are the prime divisors of the discriminant. (This includes $m < 0$.)

2.7. The supplementary symbol $(a, b/\infty)$ is defined as -1 when $a < 0$ and $b < 0$, and $+1$ otherwise. Verify that quadratic reciprocity is equivalent to

$$\prod_{p} (a, b/p) = (a, b/\infty)$$

where the indicated product is over all primes.

2.8. Hilbert's main result for his symbol is that if, for a field of discriminant d, $(n, d/p) = 1$ for all p (including ∞), then (and only then) n is equal to the norm of a number (not necessarily an integer) in $\mathbf{Q}(\sqrt{d})$. (This result will be essentially proved in Section 8.2.) For now show that this implies Legendre's theorem.

Bibliographic note

A scholarly analysis of Kummer's approach appears in Edwards (1977). An introductory discussion of genus and ring class fields appears in Cohn (1978). The classic number-theoretic background can be seen in Dickson (1938), and some additional historical perspective can be obtained from Dirichlet (1894), Hilbert (1897), Weyl (1940), and Weil (1983). For further elaboration of Remark 2.1.19, see Lagarias (1983).

3

Interpretation by rings and ideals

In this chapter we describe the splitting process in terms of algebraic concepts – namely, rings and ideals. The theory is so much a universal tool of mathematics that it will be presupposed and no effort will be made even to outline it systematically, much less to prove the main results.

We are dealing with a field $K = \mathbf{Q}(\theta)$ where θ is (say) an integer with a defining polynomial $f(x)$ in $\mathbf{Z}[x]$ that satisfies an irreducible (monic) equation. We treat the splitting of $f(x)$ (of degree m) into distinct linear factors as the assertion that m homomorphisms exist of O, the ring of integers of K onto integers $\mathbf{Z}$ mod p. This elementary fact is the key to the algebraic tools.

3.1. Rings and modules

We start with the concept of a ring R (assumed always to be commutative with a unit of one and no zero divisors). It is sometimes called an "(integral) domain." The standard definitions are as follows:

1. The nonzero elements:

$$(3.1.1) \qquad R^{\cdot} = \{\alpha : \alpha \in R, \alpha \neq 0\} = R/\{0\}.$$

2. The "set" of units:

$$(3.1.2) \qquad R^* = \{\alpha : \alpha \in R, \alpha \mid 1\}$$

that is, an element β exists in R such that $\alpha\beta = 1$.

3. The "quotient field":

$$(3.1.3) \qquad R/R^{\cdot} = \{\alpha/\beta : \alpha \in R, \beta \in R^{\cdot}\}.$$

4. Two elements whose ratio is a unit are called "associates."

A nonzero nonunit element α is called "indecomposable" exactly when there is no decomposition in R of the type $\alpha = \beta\gamma$ unless β or γ is a unit. A ring R is called a "factorial" ring, or a "unique factorization domain" (ufd), if every nonunit nonzero element α factors only into a (finite) product of indecomposable factors; that is,

$$(3.1.4a) \qquad \alpha = \kappa_1 \cdots \kappa_t \qquad (t \geq 1)$$

with the further property that the decomposition is unique. This means that for any other decomposition into indecomposables,

$$(3.1.4b) \qquad \alpha = \mu_1 \cdots \mu_s \qquad (s \geq 1)$$

there will be the "obvious match" (i.e., $t = s$) and the elements κ_i and μ_j can be renumbered to pair off as associates.

The "primes" are those nonzero nonunits π of R for which

$$(3.1.4c) \qquad \pi \mid \alpha\beta \Rightarrow \pi \mid \alpha \quad \text{or} \quad \pi \mid \beta$$

for all α, β in R. Easily, all primes are indecomposable but, conversely, all indecomposables are primes only for factorial rings. Such rings have the further property that any fraction α/β in the quotient field can be reduced so that α and β have no common primes (i.e., no primes identifiable as associates).

3.1.5. Remark. In the context of Section 2.2 we see we are dealing with the form $F_d(x, y)$ of discriminant d and this leads to the ring

$$R = \mathbf{Z}[\omega] = \{x + y\omega : x, y \in \mathbf{Z}\}$$
$$(3.1.5a) \qquad \omega = \begin{cases} (1/2)\sqrt{d} & d \equiv 0 \bmod 4 \\ (1 + \sqrt{d})/2 & d \equiv 1 \bmod 4 \end{cases}$$

with quotient field $\mathbf{Q}(\sqrt{d})$. Note that the set R is multiplicatively closed because ω is monic; that is, $\omega^2 = d/4$ or $\omega^2 = \omega - (d - 1)/4$ according as $d \equiv 0$ or $1 \bmod 4$. Then the form is defined as

$$(3.1.5b) \qquad F_d(x, y) = N\xi, \qquad \xi = x + y\omega$$

where $N\xi = \xi\xi'$, the norm over $\mathbf{Q}$. (Here ξ' is the conjugate formed by transforming $\sqrt{d} \to -\sqrt{d}$.) Also d is not assumed fundamental.

Although units and primality will be discussed more fully later, we note here that a necessary and sufficient condition for η to be a unit is that $N\eta = \pm 1$. Also the various possibilities for splitting $\alpha = \beta\gamma$ are limited by the possibilities of factoring (in $\mathbf{Z}$) the norm $N\alpha$ as $N\beta$ times $N\gamma$. (See Exercise 3.6.)

3.1.6. Lemma. *Let the ring R in equation (3.1.5a) be a factorial ring with the further property that either $d < 0$ (so all norms are positive) or if $d > 0$ then $-1 = N\eta$ for some unit η. Then the prime sets (of Section 2.2) are all equal (wfe): $P_1 = P_2 = P_3 = P_4$.*

Proof. We have only to show that

$$(3.1.6a) \qquad pz = N\xi, \qquad p \mid z \Rightarrow p = N\mu \qquad (\xi, \mu \in R)$$

so P_4 can be no smaller than P_1. To see this, start with $pz = \xi\xi'$ and separate the right-hand side into indecomposables, some subset of which $\kappa_1 \ldots \kappa_t$ must be an associate of p. Since $(p, z) = 1, t > 1$ (otherwise p is an associate of κ_1 and $N\kappa_1 = \pm p^2$). Actually the conjugate product $\kappa'_1 \cdots \kappa'_t$ is also an associate of p so $2N\kappa_1 \ldots N\kappa_t = \pm p^2$. Since no units are involved, $t = 2$ and each $N\kappa_j = \pm p$. We use a unit η, if necessary, to eliminate the minus sign and obtain $p = N\mu$. $\qquad \square$

In practice only a few fields have this property (see Exercises 3.2 and 3.3). We must therefore go beyond the role of ring properties and investigate module properties.

If S is a subring of R, we define $\mathfrak{M}$, an "S-module," as a subset of R closed under addition, subtraction, and multiplication by S. If S is not specifically designated, it is taken as $\mathbf{Z}$. Trivially, R is an R-module as well as an S-module for each S. We call $\mathfrak{M}$ a "finite" S-module if every element $\alpha \in \mathfrak{M}$ can be expressed in terms of a "finite (fixed) basis" $\beta_1, \ldots, \beta_t$ in R and a (variable) set $\sigma_1, \ldots, \sigma_t$ in S; that is,

$$(3.1.7a) \qquad \alpha = \sigma_1\beta_1 + \cdots + \sigma_t\beta_t, \qquad \sigma_i \in S, \beta_i \in R.$$

This is written

$$(3.1.7b) \qquad \mathfrak{M} = [\beta_1, \ldots, \beta_t]S$$

with S omitted if $\mathbf{Z}$ is understood. Thus, equation (3.1.5a) could be written as

$$R = [1, \omega].$$

The basis is called "minimal" if $\alpha = 0$ only when all $\sigma_i = 0$. For a minimal basis, t is the S-rank of M. (Equivalently, the representation of α is unique in σ_i for a minimal basis β_i.) If $K \supset k$ for two finite-degree fields, then $K (= R)$ is a k-$(= S$-)module with a basis of m elements where $m = [K : k]$. In fact, if $K = k(\xi)$, a basis is given by

$$(3.1.7c) \qquad K = [1, \xi, \ldots, \xi^{m-1}]\mathbf{Q}.$$

In practice, to recognize equivalent bases (particularly nonminimal bases) can be difficult. For instance, if $\mathfrak{M}$ is an S-module, then

$$\mathfrak{M} = [\alpha, \beta, \ldots]S.$$

and equivalent bases are generated by operations such as the following:

$$(3.1.8) \quad \begin{cases} \alpha \to \beta, \quad \beta \to \alpha \\ \alpha \to \alpha + \sigma\beta, \quad \sigma \in S \\ \text{eliminate } \beta \text{ if } \alpha \mid \beta \text{ (special case } \beta = 0). \end{cases}$$

We next define "integers" in R, but this is done relatively to a subring S (usually taken as $\mathbf{Z}$). We first consider the quotient field $K = R/R^{\cdot}$ as a ring with R as a subring. Then for $\xi \in K$, $R[\xi]$ is a subring of K leading to the inclusion:

$$(3.1.9) \qquad K \supseteq R[\xi] \supseteq R \supseteq S.$$

We say that ξ is "S-integral" over R exactly when $R[\xi]$ is a finite S-module. An equivalent form is that ξ satisfies a monic equation over S:

$$(3.1.10) \qquad \xi^m + \alpha_1\xi^{m-1} + \cdots + \alpha_m = 0 \qquad (\alpha_i \in S).$$

We call T, the set of S-integral elements of K, the "S-(integral)-closure" or the "S-integers" in K for R. Here T is written

$$(3.1.11) \quad T = S\text{-clos } R = \{\xi : \xi \text{ satisfies equation (3.1.10)}\}.$$

Furthermore, T is a ring. (This is less obvious and is omitted.) Clearly, S-clos $R \supseteq S$ (since $\xi - \xi$ is monic), but more important, the closure is its own closure (it creates no more integers than S):

$$(3.1.12) \qquad T = T\text{-clos } R.$$

A ring that is its closure is called "integrally closed."

We shall usually use $\mathbf{Z}$ as the ring S so all closures are $\mathbf{Z}$-closures. If $K = \mathbf{Q}(\theta)$ where θ is of degree n $(= [K : k])$, then the ring of integers in K is the $\mathbf{Z}$-closure denoted by

$$(3.1.13a) \qquad O = \mathbf{Z}\text{-clos } K$$

denoted more "explicitly" by

(3.1.13b) $O = \{\theta \in K : \theta^m + a_1\theta^{m-1} + \cdots + a_m = 0, \quad a_i \in \mathbf{Z}\}.$

Thus, in addition, for the integer ring, $O = O\text{-clos } K$. We later see that O always has a minimal basis over $\mathbf{Z}$. This is not trivial in the following sense: If we have an extension $K' \supset K$, its integers O' would have a minimal basis over $\mathbf{Z}$ but not always over O (only a finite basis over O). This is important later (see Section 6.1).

3.1.14. Theorem. *A factorial ring must be integrally closed.*

Proof. Otherwise some $\xi = \alpha/\beta$ $(\alpha \in R, \beta \in R^{\cdot})$ satisfies equation (3.1.10) for $\alpha_i \in R$. This means an indecomposable ω (in R) divides β but not α. We now have a contradiction in the form

(3.1.14a) $\alpha^m/\beta = -(\alpha_1\alpha^{m-1} + \alpha_2\alpha^{m-2}\beta + \cdots + \alpha_{m-1}\beta^{m-1}).$

(The left side cannot be an integer while the right is.)   □

We now formalize the notation for the field $K = \mathbf{Q}(\sqrt{D})$ where D is square-free. The integers are seen to be the $\mathbf{Z}$-module:

$$O = [1, \omega] = \{x + y\omega : x, y \in \mathbf{Z}\}$$
(3.1.15) $\omega = \begin{matrix} \sqrt{D} & D \not\equiv 1 \bmod 4 \\ (1 + \sqrt{D})/2 & D \equiv 1 \bmod 4 \end{matrix}$

corresponding to forms $F_d(x, y)$ where d $(= d_1)$ is the fundamental discriminant (see Section 2.2):

(3.1.16) $d = \begin{cases} 4D & D \not\equiv 1 \bmod 4 \\ D & D \equiv 1 \bmod 4. \end{cases}$

We also call d the "field discriminant." It is an invariant (in a sense discussed in the next chapter); for instance, if D is changed to m^2D, then D still generates the same field characterized by the same discriminant d. (See Exercise 3.1.) If the double condition on D seems troublesome, we can write

(3.1.17) $O = \mathbf{Z}[\omega] = [1, \omega], \quad \omega = (d + \sqrt{d})/2.$

3.1.18. Remark. For the nonfundamental discriminant $d_f = f^2d$ $(f \geq 1)$, the values of $F_{d_f}(x, y)$ for $x, y \in \mathbf{Z}$ are norms [see equations (2.2.6a, b)] in the ring or module

(3.1.18a) $$O_f = \mathbf{Z}[f\omega] = [1, f\omega].$$

Clearly, if $f > 1$, then O_f lacks ω and is therefore not integrally closed.

3.1.19. Theorem. *The prime p (odd or even) splits the defining equation for* ω *[in equation (3.1.15)] exactly when (d/p) = 1 (Kronecker's symbol). All splitting primes satisfy* $p = F_d(x, y)$ *when O is a factorial ring. (Here we take d to be fundamental only.)*

The easiest condition for inducing a ring R to be factorial follows: Let the field $K = R/R^{\cdot}$ have a norm operation N. Then we call R "euclidean" exactly when, for every α/β in K, there exists a ξ in R such that

(3.1.20) $$|N(\xi - \alpha/\beta)| < 1.$$

3.1.21. Theorem. *For a euclidean ring R a greatest common divisor (gcd) exists; that is, for each* $\alpha, \beta \in R$, γ [$= gcd (\alpha, \beta)$] *exists that divides* α *and* β *(in R) and is at the same time divisible by every divisor (in R) of* α *and* β. *Also* $\gamma = \alpha\xi + \beta\eta$ *for suitable* ξ *and* η *(in R).*

Proof. From expression (3.1.20), $|N(\beta)| > |N(\alpha - \beta\xi)|$. Thus, if we define $\alpha = \alpha_1$ and $\beta = \alpha_2$, we can set up a chain of α_t inductively by defining ξ_t so that $|N(\alpha_t/\alpha_{t+1} - \xi_t)| < 1$ and

(3.1.21a) $$\alpha_t - \xi_t\alpha_{t+1} = \alpha_{t+2}.$$

This creates a chain of α_t of decreasing norm, terminating when some $\alpha_{T+1} = 0$. But the common divisors of (α_t, α_{t+1}) are all the same as those of $(\alpha_{t+1}, \alpha_{t+2})$ by equation (3.1.21a). This chain ends with $(\alpha_T, 0)$ or simply α_T, the gcd of α and β. □

3.1.22. Corollary. *A euclidean ring is factorial. Consequently the gcd* (α, β) *is deducible from the factorizations of* α *and* β, *and hence the gcd is unique except to within associates (unit factors).*

Proof. If π is indecomposable, we need only show that if $\pi \mid \alpha\beta$, then $\pi \nmid \alpha$ or $\pi \mid \beta$. This sets up a cancellation process for matching any two factorizations of a number in R into indecomposables (primes).

Suppose $\pi \nmid \alpha$, then gcd $(\pi, \alpha) = 1 = \pi\xi + \alpha\eta$ (for $\xi, \eta \in R$). Hence, $\beta = \beta\pi\xi + \beta\alpha\eta$, so π divides β (since it divides $\alpha\beta$). □

3.1.23. Remark. A field is called "euclidean," "unique factorization," "factorial," and so on according to O, its ring of integers.

3.1.24. Remark. In elementary number theory we prove **Z** to be euclidean, and hence factorial, by choosing the "round-off" value of x in **Z** for a fraction a/b by $|a/b - x| \leq \frac{1}{2}$. The round-off method works for surprisingly many fields (see Exercise 3.5). There are even many more situations in which a field is euclidian but the choice of the value of ξ in inequality (3.1.20) is not at all an obvious round-off. Euclidean fields remain a lively area of investigation.

In polynomial ring theory, with $R = F[x]$ for F a field, the euclidean property likewise comes from division. Instead of the norm, we define $\deg A(x)$ as the degree of the polynomial $A(x) \in R$. (Thus, if $a \in R$, $\deg a = 0$, except $\deg 0 = -\infty$.) If $B(x) \in R^{\cdot}$, then $A(x) = B(x)Q(x) + R(x)$ for quotient $Q(x)$ and remainder $R(x)$. The euclidean property is that $\deg R(x) < \deg B(x)$.

3.2. Ideals and divisors

We define an "ideal" $\mathfrak{I}$ in a ring R as an R-module. Thus, axiomatically, if $\alpha, \beta \in \mathfrak{I}$ and $\rho \in R$, then

(3.2.1a) $\qquad\qquad \alpha \pm \beta \in \mathfrak{I} \qquad (\mathfrak{I} \pm \mathfrak{I} \subseteq \mathfrak{I})$

(3.2.1b) $\qquad\qquad \rho\alpha \in \mathfrak{I} \qquad (\mathfrak{I}R \subseteq \mathfrak{I})$.

So, if $\{\alpha_1, \ldots, \alpha_r\}$ is any set of elements of R, it can serve as an ideal basis (of $\mathfrak{I}$ as an R-module):

$$
(3.2.2) \qquad
\begin{aligned}
\mathfrak{I} &= [\alpha_1, \ldots, \alpha_r]R = (\alpha_1, \ldots, \alpha_r) \\
&= \{\alpha_1\xi_1 + \cdots + \alpha_r\xi_r\}, \qquad \xi_i \in R.
\end{aligned}
$$

Here the symbol "(...)" denotes the ideal, as an abbreviation for "[...]R." We assume the basis to be finite, although it may be infinite as long as only a finite number of elements are used at a time. The motivation for the definition is that properties (3.2.1a, b) characterize the set of elements of R that are divisible by some fixed element $\gamma \in R$. Sometimes an ideal is called a "divisor." By convention, an ideal that consists of only $\{0\}$ (or an empty set) is excluded (although it satisfies the axioms).

We define a "principal" ideal (α) as

(3.2.3) $\qquad\qquad \mathfrak{I} = (\alpha) = \alpha R = \{\alpha\xi : \xi \in R\}$.

Clearly, $(\beta) \subseteq (\alpha) \leftrightarrow \beta \in (\alpha) \leftrightarrow \beta = \alpha\xi \leftrightarrow (\alpha) \mid (\beta)$. Thus, $(\alpha) = (\beta)$ if and only if α and β are associates. We call R a "principal ideal domain" (ring) or a "pid" when all ideals are principal. In particular, $R = (1)$. Generally, when the meaning is clear, the principal ideal (α) will be denoted by α, so we write $R = 1$.

The sum of two ideals $\mathfrak{I}_1 + \mathfrak{I}_2$ is the set of sums $\{\alpha_1 + \alpha_2\}$ for all $\alpha_1 \in \mathfrak{I}_1$ and $\alpha_2 \in \mathfrak{I}_2$. The product of two ideals is not the corresponding set of products but the ideal generated by that set. In principle, this is an infinite basis, but when each ideal has a finite basis, for example,

(3.2.4a) $$\mathfrak{a} = (\alpha_1, \ldots, \alpha_r), \qquad \mathfrak{b} = (\beta_1, \ldots, \beta_s)$$

then the product has a basis (of rs elements)

(3.2.4b) $$\mathfrak{a}\mathfrak{b} = (\alpha_1\beta_1, \ldots, \alpha_r\beta_s).$$

The type of product (3.2.4b) would define a product of any S-module (not just for $S = R$). It is called a "module product." Finally, it is easily seen that $\mathfrak{a}\mathfrak{b} \subseteq \mathfrak{a}$ and $\mathfrak{a}\mathfrak{b} \subseteq \mathfrak{b}$. In a converse manner, Dedekind defined "division" by saying "to divide is to contain"; that is,

(3.2.5) $$\mathfrak{a} \mid \mathfrak{b} \leftrightarrow \mathfrak{a} \supseteq \mathfrak{b}.$$

It then becomes necessary to show (as is true for ideals in number theory) that if $\mathfrak{a} \supseteq \mathfrak{b}$, then an ideal $\mathfrak{c}$ exists in R so that $\mathfrak{a}\mathfrak{c} = \mathfrak{b}$. An equivalent condition for $\mathfrak{a} \mid \mathfrak{b}$ is $\mathfrak{a} + \mathfrak{b} = \mathfrak{a}$.

The special concept of division introduces a new definition. An "indecomposable" ideal $\mathfrak{m}$ in $\mathfrak{R}$ is now called "maximal" when

(3.2.6a) $$R \supseteq \mathfrak{a} \supseteq \mathfrak{m} \Rightarrow \mathfrak{a} = R \quad \text{or} \quad \mathfrak{a} = \mathfrak{m}.$$

An ideal $\mathfrak{p}$ is called "prime" when

(3.2.6b) $$\mathfrak{p} \mid \mathfrak{a}\mathfrak{b} \Rightarrow \mathfrak{p} \mid \mathfrak{a} \quad \text{or} \quad \mathfrak{p} \mid \mathfrak{b}.$$

It follows that all maximal ideals are prime (not conversely). The gcd ideal becomes the "maximal common containing" ideal, now a crucial concept. Any ideal is the gcd of its basis. As an immediate result, a euclidean ring is automatically a principal ideal ring (since the gcd is a single element).

We can consider the ideals in a ring R to be a semigroup with a unit of R. We then consider factorization of an arbitrary (nonzero) ideal $\mathfrak{I}$. There may or may not be only a finite number of factors of $\mathfrak{I}$ ($= \mathfrak{A}\mathfrak{B}$) for other ideals $\mathfrak{A}$, $\mathfrak{B}$, and even so, there may be different ways to decompose $\mathfrak{I}$ into indecomposable ideal factors. None of this is certain without further assumptions. Indeed, unique factorization (into finite strings of prime ideals) is valid if and only if R satisfies three axioms of a so-called Dedekind ring:

1. R is integrally closed in its quotient field.
2. Prime ideals are maximal.
3. There is a finite basis for all ideals in R.

These axioms are independent (but have many interesting equivalent forms not discussed here). They are seen immediately to be valid for euclidean rings. Furthermore they hold up under a finite extension of K (the quotient field of R) to K_1, with $R_1 = R$-clos K_1 (R_1 is the set of integers in K_1). Then R_1 is a Dedekind ring when R is. Thus, Dedekind rings can be propagated (by extensions) starting with elementary, well-known (euclidean) rings:

1. $R = \mathbf{Z}$, extensions produce (algebraic) "number fields," and
2. $R = F[z]$, for F a field, extensions produce (algebraic) "function fields" of one variable.

The most familiar example of a useful non-Dedekind ring involves algebraic functions of several variables, $R = F[x, y]$. This is a factorial ring that is not a Dedekind ring. It has a prime ideal (x) that is not maximal, $(x) \subset (x, y)$. In a Dedekind ring, however, the factorial property ensures that the ring is a principal ideal ring, and conversely. When a Dedekind ring is factorial, then all factorizations into ideal correspond directly to factorizations in the ring (if an ideal is identified with the associates of a generating element). (See Exercise 3.8.)

A Dedekind ring R has a unique factorization into maximal (prime) ideals. The ideals of R are called "integral ideals" with an obvious semigroup structure (identity is R). To obtain an "ideal group" structure, we introduce "fractional ideals." These are ideals (i.e., R-modules) in the quotient field K, characterized as further having a bounded denominator. Thus, if $\mathfrak{I}$ is an integral ideal, the most general fractional ideal is

$$(3.2.7) \qquad \mathfrak{I}_0 = \mathfrak{I}/\alpha = \{\xi/\alpha : \xi \in \mathfrak{I}\}, \qquad \alpha \in R^{\cdot}.$$

Thus, any ideal quotient $\mathfrak{a}/\mathfrak{b}$ ($\mathfrak{a}, \mathfrak{b} \neq 0$) is of the form (3.2.7). Moreover, every integral (or fractional) ideal $\mathfrak{a}$ has a two-element basis α, β in R (or K):

$$(3.2.8) \qquad \mathfrak{a} = (\alpha, \beta) = (\alpha) + (\beta) = \alpha R + \beta R$$

where α is chosen arbitrarily in $\mathfrak{a}$. In fact, we can choose any ideal $\mathfrak{I}$ $\subseteq \mathfrak{a}$ and write $\mathfrak{a} = \mathfrak{I} + (\beta)$.

The ideal group is now broken up into equivalence classes. Two ideals are called "equivalent" if their ratio is a principal (possibly) fractional ideal. Equivalence of ideals $\Im$ and $\Im'$ is denoted by $\Im \frown \Im'$. The class of ideals equivalent to $\Im$ is denoted by $[\Im]$, so equivalence would also be written $[\Im] = [\Im']$. It is also part of the theory that every ideal class is represented by an integral ideal relatively prime to any preassigned ideal.

If we consider K/k an extension, we have an integer ring $O_K \supseteq O_k$ in each field. A prime factorization in k will generally have to be further factored to produce primes in K. Each ideal $\Im$ in O_k will correspond to an imbedded ideal $\Im O_K$ in O_K (defined by a module product because both $\Im$ and O_K are O_k-modules). The ideal $\Im O_k$ will be denoted by $\Im$ (even after imbedding in O_K) as long as the context is clear.

If K/k is normal, the action of the Galois group will produce $m = [K:k]$ conjugates to any ideal $\mathfrak{A}$ in O_K,

$$(3.2.9) \qquad \mathfrak{A}, \mathfrak{A}', \mathfrak{A}'', \ldots, \mathfrak{A}^{(m-1)}.$$

Their product defines a "relative norm":

$$(3.2.10) \qquad N_{K/k}\mathfrak{A} = \mathfrak{A}\mathfrak{A}'\mathfrak{A}'' \cdots \mathfrak{A}^{(m-1)}.$$

This is an ideal in O_K, but it can be identified with one imbedded from O_k, so we call the norm an ideal of O_k. A relative norm would be defined similarly for any integer A, and of course if $\mathfrak{A}$ is principal, the concepts of relative norm would agree.

In particular, for $K/\mathbf{Q}$, there is the "absolute norm" of an ideal $\mathfrak{A}$ of O_K generated by a positive integer called $N\mathfrak{A}$ (or $N[\mathfrak{A}]$). There is an alternate definition:

$$(3.2.11) \qquad N[\mathfrak{A}] = [O_K : \mathfrak{A}],$$

the cardinality of the residue classes of O_K modulo $\mathfrak{A}$. As before, there is also the absolute norm of B, an integer of K, but NB might be a negative integer, so $N[(B)] = |NB|$. Technically, this agrees with equation (3.2.10):

$$(3.2.12) \qquad N_{K/\mathbf{Q}}(B) = (N[(B)]) = N[(B)]\mathbf{Z}.$$

The norms are multiplicative under any definition; that is,

$$(3.2.13) \qquad N[\mathfrak{A}\mathfrak{B}] = N[\mathfrak{A}]N[\mathfrak{B}].$$

The definition $N[\mathfrak{M}] = [R : \mathfrak{M}]$ could be likewise extended to any module $\mathfrak{M}$ in a ring R, but the multiplicativity (3.2.13) is not generally true except for ideals in a Dedekind ring.

If $\mathfrak{P}$ is a prime ideal in K and if it divides $\mathfrak{p}$ in k, then

(3.2.14a) $$N_{K/k}\,\mathfrak{P} = \mathfrak{p}^f$$

and, in particular, for $K/\mathbf{Q}$,

(3.2.14b) $$N\mathfrak{P} = p^f.$$

Here f is called the "relative (or absolute) degree" of $\mathfrak{P}$ (respectively). This is an assertion of an algebraic property of general abstract significance, that since $\mathfrak{P}$ is maximal, then (say) $O_K/\mathfrak{P}$ is a field of a finite number of elements, with $\mathbf{Z}/p$ as a subfield. If K is not a normal field, it still has a normal closure L, and the action of the Galois group takes place on the imbeddings in L of the ideals of K.

3.3. Ideals as homomorphisms

The special relevance of ideals to the splitting concept comes from the fact that each ideal $\mathfrak{I}$ in R (including the zero ideal) corresponds uniquely to a homomorphic mapping h of the ring R onto an image (with kernel $\mathfrak{I}$) definable by

(3.3.1) $$h : R \rightarrow R/\mathfrak{I}.$$

Here h identifies exactly the elements of R congruent mod $\mathfrak{I}$. [Thus, $h(\alpha) = h(\beta) \leftrightarrow \alpha - \beta \in \mathfrak{I}$.] Conversely, every homomorphism h leads to an ideal; that is, $\mathfrak{I} = \{\alpha : h(\alpha) = 0\}$. The axioms (3.2.1a, b) are precisely equivalent to this correspondence.

First, take the case where the integers O of K are generated by a single integer θ [root of the monic equation $f(x) = 0$], or

(3.3.2) $$O = Z[\theta].$$

We can now explain the splitting of K by p in terms of ideals. Here the congruence $f(x) = 0$ mod p has m roots ($m = [K : \mathbf{Q}]$), namely, $x_1, \ldots, x_m$, which are all different when $f(s)$ is split by p. Hence, the m correspondences follow:

(3.3.3a) $$\theta \rightarrow x_i, \qquad i = 1, \ldots, m.$$

These are m different homomorphisms of R onto $\mathbf{Z}/p\mathbf{Z}$. Each of these m correspondences is given by taking an arbitrary element $g(\theta)$ in O and setting

(3.3.3b) $$g(\theta) \rightarrow g(x_i) \text{ mod } p.$$

Each such correspondence must also correspond to an ideal $\mathfrak{P}_i$ in O, which contains pO because any integral multiple of $\mathfrak{P}$ goes to O in relation (3.3.3b). Of course, this means that $\mathfrak{P}_i$ divides pO. (In a later result, Theorem 3.3.5 will assure us that these ideals are a complete factorization of pO.) By relation (3.3.3b), $N\mathfrak{P}_i = [O : \mathfrak{P}_i] = |\mathbf{Z}/(p)| = p$.

We will see in Chapter 4 that not every ring O can be expressed as $\mathbf{Z}[\theta]$ for a suitable θ. The next result is more general.

3.3.4. Theorem. *Let $K = \mathbf{Q}(\theta)$ be a normal extension with θ taken (for convenience) as an integer defined by the monic polynomial $f(\theta) = 0$. Let the discriminant be*

$$(3.3.4a) \qquad d = d(\theta) = \prod_{t > u} (\theta_t - \theta_u)^2.$$

Then if $p \nmid d$, p splits $f(x)$ exactly when its factorization in O is given by distinct primes of degree one as follows:

$$(3.3.4b) \qquad pO = \mathfrak{P}_1 \cdots \mathfrak{P}_m.$$

Although we do not supply proofs in this chapter, it may be noted that if the factorization (3.3.4b) occurs, then there are m distinct homomorphisms (3.3.3a) and hence m distinct factors $x - x_i$ in $f(x)$. So prime factorization implies splitting. The converse was noted earlier. Anyhow, this result is part of the more general Theorem 3.3.5 (of Dedekind):

3.3.5. Theorem. *Let $K = \mathbf{Q}(\theta)$ be an extension field not necessarily normal but generated by an integer that satisfies the monic equation*

$$(3.3.5a) \qquad f(x) = x^m + a_1 x^{m-1} + \cdots + a_m = 0.$$

Then $\mathbf{Z}[\theta]$ is generally only a submodule of O, the integer ring of K, of index $j = [O : \mathbf{Z}[\theta]]$. Assume $(p, j) = 1$. Then if we factor

$$(3.3.5b) \qquad f(x) \equiv f_1(x)^{e_1} \cdots f_g(x)^{e_g} \bmod p$$

into polynomials $f_i(x)$ irreducible mod p and distinct, this factorization simulates the prime factorization of pO,

$$(3.3.5c) \qquad p = \mathfrak{P}_1^{e_1} \cdots \mathfrak{P}_g^{e_g}.$$

If the polynomials $f_i(x)$ are of degree f_i, then this is the degree of $\mathfrak{P}_i$, which can now be written in the form

(3.3.5d) $$\mathfrak{P}_i = (p, f_i(\theta)).$$

[*So when the degree is one, the ideal $\mathfrak{P}$ takes the form $(p, \theta - a)$ for a in* **Z**.] *The degrees satisfy*

(3.3.5e) $$m = \sum_{i=1}^{g} e_i f_i.$$

If one (or more) $e_i > 1$, then the prime p is said to be "ramified" over (all such) $\mathfrak{P}_i$ with the ramification "order" e_i. If $g = 1$ and $e_1 = 1$, then by equation (3.3.5c) $f_1 = m$. No further factorization occurs $(p = \mathfrak{P})$ and the prime p is said to be "inert." If all $e_i = f_i = 1$ (so $m = g$), then p is said to "split." These same definitions are said to apply to a factorial ring O when the prime p is factored in O. They also apply when a prime $\mathfrak{P}$ in K is factored in primes over a larger field L.

Note that the homomorphism $O/\mathfrak{P}_i$ represents a finite field of p^{f_i} elements. If the field is normal, the degrees f_i of all the polynomials $f_i(x)$ are the same as well as the ramification orders e_i, so

(3.3.6) $$m = efg.$$

3.3.7. Illustration. Let $O = \mathbf{Z}[i] = \{a + bi : a, b \in \mathbf{Z}\}$. Then $p = (2 + i)$ is the kernel of the homomorphism $h(i) \equiv -2 \mod 5$, $h(a + bi) \equiv a - 2b \mod 5$. So $f = 1$ and $h(O) = \mathbf{Z}/5\mathbf{Z}$, a field of five elements. But $p = (3)$ is the kernel of the homomorphism $h(a + bi) \equiv a + bi \mod 3$; that is, $h(i)$ has no image in **Z** mod 3. So now $f = 2$ and $h(O) = \mathbf{Z}/3\mathbf{Z} \oplus i\mathbf{Z}/3\mathbf{Z}$, a field of nine elements.

As an incidental result of Theorem 3.3.5, if $p \nmid [O : \mathbf{Z}[\theta]]$, then p ramifies only when $p \mid d(\theta)$. This follows from the fact that $f(x)$ and $f'(x)$ are relatively prime when factored in **Z** mod p unless $p \mid d(\theta)$.

Exercises

3.1. Verify that the integers of $\mathbf{Q}(\sqrt{D}) = K$ are those of equation (3.1.15). Note that the general element of K is $\xi = (a + b\sqrt{D})/c$, and show which values of a, b, c in **Z** (without common factors) permit ξ to be defined by a monic equation.

3.2. Verify that if a quadratic field has discriminant $d < 0$, then $p = F_d(x, y)$ is not solvable if $p < |d/4|$. Thus, if $(d/p) = 1$ for such a p, then the integers of $\mathbf{Q}(\sqrt{d})$ are not factorial (not a ufd).

3.3. Verify for the quadratic field $Q(\sqrt{D})$ with $D \not\equiv 1 \bmod 4$, $D \neq \pm 2$, and $d = 4D$ that $P_1 \neq P_2$ [(see equations (2.2.7)]. Some of these fields still have factorial integer domains (as in Exercise 3.5).

3.4. Verify that in a quadratic field if $(d/p) = 1$ or 0 implies that $p = \pm N(\xi)$ for some integer ξ, then the integer ring is a ufd.

3.5. Verify that for $d = -3, -4, -7, -8$, and $d = 5, 8, 12, 13$, the integer ring of $Q(\sqrt{d})$ is euclidean. Note that for rationals $|x| \leq \frac{1}{2}$ and $|y| \leq \frac{1}{2}$, $|N((x + y\sqrt{d})/2)| \leq |(d-1)/16|$, and so on.

3.6. Show that in the ring $Z[\sqrt{-6}]$ the two decompositions $6 = 2 \cdot 3$ and $6 = -(\sqrt{-6})(\sqrt{-6})$ yield indecomposable factors that cannot be paired as associates. Verify that 2, 3, and $\sqrt{-6}$ are not primes. On the contrary, in the ring $Z[\sqrt{6}]$ the two decompositions $6 = 2 \cdot 3$ and $6 = (\sqrt{6})(\sqrt{6})$ can be reconciled by the factorization of 2, 3, and $\sqrt{6}$ into indecomposables that now are primes. The factors have the form $2 = (2 + \sqrt{6})^2\eta$ (for η a unit); likewise, $3 = (3 + \sqrt{6})^2\eta$ and $\sqrt{6} = (2 + \sqrt{6})(3 + \sqrt{6})\eta$. (Note that $|N\alpha| = 2$ or 3 is solvable in one ring but not in the other.)

This example illustrates an interesting historical point. Kummer introduced "ideal" numbers (i.e., numbers outside the ring R) that consummate unique factorization. For instance, in $R = Z[\sqrt{-6}]$, if we had $\sqrt{-6} = \sqrt{2} \cdot \sqrt{-3}$, we would be finished. We now want the "ideal" number $\sqrt{2}$, not in R ($\sqrt{-3}$ enters as $\sqrt{-6}/\sqrt{2}$). Dedekind kept within the ring R by defining the ideal $\mathfrak{I}$ as those elements ξ of R that are divisible by $\sqrt{2}$ in the sense that $\xi/\sqrt{2}$ satisfies a monic equation. Verify that $\mathfrak{I} = (2, \sqrt{-6}) = \{x + y\sqrt{-6} : x \text{ even}\}$. It was originally the role of class fields to provide such unique factorization, but the meanings changed later on (see Chapter 7). [We chose $\sqrt{2}$ rather than $\sqrt{-2}$ so that the enlarged field would be $Q(2, \sqrt{-3})$, a genus field as in Chapter 2.]

3.7. Verify that when d is the discriminant of the field $K = Q(\sqrt{D})$, the primes p that factor in O are as follows:

a. $(d/p) = 0$ (p odd) $\Rightarrow p = \mathfrak{p}^2$, $\mathfrak{p} = (p, \sqrt{D})$

b. $(d/p) = 1$ (p odd) $\Rightarrow p = \mathfrak{p}\mathfrak{p}'$, $\mathfrak{p} = (p, x + \sqrt{D})$, $\mathfrak{p}' = (p, x - \sqrt{D})$ where $x^2 \equiv D^2 \bmod p$

c. $(d/2) = 0$ (D even) $\Rightarrow 2 = \mathfrak{p}^2$, $\mathfrak{p} = (2, \sqrt{D})$

d. $(d/2) = 0$ ($D \equiv 3 \bmod 4$) $\Rightarrow 2 = \mathfrak{p}^2$, $\mathfrak{p} = (2, 1 + \sqrt{D})$

e. $(d/2) = 1$ ($D \equiv 1 \bmod 8$) $\Rightarrow 2 = \mathfrak{p}\mathfrak{p}'$, $\mathfrak{p} = (2, (1 + \sqrt{D})/2)$, $\mathfrak{p}' = (2, (1 - \sqrt{D})/2)$

f. In all other cases, for example, $(d/p) = -1$ (p odd or 2), p is inert. In cases a–e, show that the factors multiply out to (p).

3.8. A Dedekind ring has the property that if the ideals $\mathfrak{a}$ and $\mathfrak{q}$ are given, then $\beta \in \mathfrak{a}$ such that $(\beta) = \mathfrak{a}\mathfrak{c}$ for $(\mathfrak{q}, \mathfrak{c}) = 1$. Verify that β is constructed as follows: Write $\mathfrak{a}' = \mathfrak{a} \, \Pi \, \mathfrak{q}_i$ for the primes $\mathfrak{q}_i \mid \mathfrak{q}$, and $\mathfrak{a}'_i = \mathfrak{a}'/\mathfrak{q}_i$ $(\supset \mathfrak{a}')$. Then we define β_i as an element in $\mathfrak{a}'_i$ but not in $\mathfrak{a}'$, and $\beta = \Sigma \, \beta_i$. Deduce that if $\mathfrak{a} \supseteq \mathfrak{m}$, then some $\beta \in \mathfrak{a}$ satisfies $\mathfrak{a} = \mathfrak{m} + (\beta)$.

3.9. A Dedekind ring is a pid if and only if it is factorial. *Hint:* The hard step is to show for R factorial that the indecomposable (prime) elements generate all the prime ideals. Otherwise there would be a nonprincipal prime ideal $\mathfrak{p}$ for which we could write $\mathfrak{p}\mathfrak{p}^* = (\pi)$, for π prime. Then $\mathfrak{p} \sim \mathfrak{q}$, $\mathfrak{p}^* \sim \mathfrak{q}^*$ for $(\mathfrak{q}, \pi) = (\mathfrak{q}^*, \pi) = 1$. So $\pi \mid (\mathfrak{p}\mathfrak{q}^*)$ $(\mathfrak{q}\mathfrak{p}^*)$. Both factors are principal ideals; this leads to a contradiction.

Bibliographic note

We owe the classical ring theory to Dedekind (1871), and it was further conceptualized by Noether (1921). Classic references are Van der Waerden (1950) and Jacobson (1974). H. W. Lenstra (1975) has done some recent work on the euclidean algorithm (see Remark 3.1.24).

4

Finite invariants of a field

It is well recognized that ideal theory was at least partly responsible for opening the Pandora's box of the infinite. The idea of replacing a number (as divisor) by an ideal (an infinite set) worked so well that it encouraged this kind of thinking of infinite aggregates as the most basic concept. Thus, fields were viewed as infinite sets that satisfy certain axioms, and not as the range of elements constructed (say) as polynomials in their generators.

At the same time, anything useful that is done with algebraic numbers refers to a "finite set of answers," for example, the values (or even the existence) of integers x, y for which a form $F(x, y) = p$, a prime, or the value of an algebraic number θ for which $\mathbf{Q}(\theta)$ is the desired class field. To begin with, there are certain finite invariants that distinguish (if not characterize) the given field K:

1. Integers O are imbedded in a substructure of K measured by the discriminant d.
2. Units are imbedded in a substructure of O measured by the regulator R.
3. Ideals are embedded in a group structure measured by the class number h.

These three structures are remarkable in that they are all discrete in the topological sense, and the three constants d, R, and h are related. The relation is through the zeta-function, which is part of hard-core analysis ("infinitesimal processes"). We need this relationship because it presents an upper limit to the degree of the class field for a set of primes, but as usual we only outline the theory.

For many years, algebraic number theory was beset by the desire to free the "pure" algebraic results from the "applied" analytic theory.

The more current theory seems headed toward proving that this is unnecessary because the algebraic and analytic aspects have a common origin and never were really different.

4.1. Integers and discriminants

Let θ be an algebraic integer that satisfies an irreducible equation of degree m:

(4.1.1) $$\theta^m + a_1\theta^{m-1} + \cdots + a_m = 0.$$

We define the discriminant of θ as the rational integer

(4.1.2) $$d(\theta) = \prod_{i>j} (\theta^{(i)} - \theta^{(j)})^2$$

where $\theta^{(i)}$ represents the conjugates of θ ($= \theta^{(1)}$). For technical reasons apparent later (see Remark 4.1.18), we do not make discriminants positive [by using absolute values in equation (4.1.2), for instance]. Each determines a module:

(4.1.3) $$\mathbf{Z}[\theta] = [1, \theta, \ldots, \theta^{m-1}].$$

It would be desirable to say that for any field $K = \mathbf{Q}(\theta)$, some generating θ can be chosen so that the range of integers O is merely $\mathbf{Z}[\theta]$. Such a ring O is called "monogenic," but if $m > 2$, not every O has this property. The most we know is that $[O : \mathbf{Z}[\theta]]$ is finite. Indeed if α is an arbitrary integer in O and $\alpha^{(i)}$ is the conjugate, we can always set

(4.1.4) $$\alpha^{(j)} = \sum_{k=0}^{m-1} r_k\theta^{(j)k} = r_0 + \cdots + r_{m-1}\theta^{(j)m-1}$$

for $r_0, \ldots, r_{m-1}$ rational. The denominators of the r_k can be expressed as the determinant

(4.1.5) $$D = \begin{vmatrix} 1 & \theta^{(1)} \cdot & \cdot \cdot \theta^{(1)m-1} \\ & \cdot & \cdot \\ & \cdot & \cdot \\ & \cdot & \cdot \\ 1 & \theta^{(m)} \cdot & \cdot \cdot \theta^{(m)m-1} \end{vmatrix}$$

with numerators that are algebraic integers. The denominators can be taken as $D^2 = d(\theta)$ (Vandermonde's theorem) and the numerators are

(4.1.6) $$\mathbf{Z}[\theta]/d(\theta) \ (\supseteq O \supseteq \mathbf{Z}[\theta])$$

and the finite generation of O is established (almost). We need only test each of the finite sets

(4.1.6a)
$$\xi = \sum_{k=0}^{m-1} b_k \theta^k / |d(\theta)|$$

for all $0 \le b_K \le |d(\theta)|$ (surely too many in practice!) to see which of them are integers. For such $\xi = \xi_t$, where $1 \le t \le T$,

(4.1.7)
$$O = \mathbf{Z}[\theta] + \xi_1 \mathbf{Z} + \cdots + \xi_T \mathbf{Z}.$$

This is not a direct sum (minimal basis). To obtain one, we note the following results:

4.1.8. Theorem. *A finitely generated $\mathbf{Z}$-module has a minimal basis.*

4.1.9. Theorem (Hermite–Smith). *If O has the minimal basis*

(4.1.9a)
$$O = [\alpha_1, \ldots, \alpha_m]$$

then any submodule $\mathfrak{M}$ of O with the same $\mathbf{Z}$-rank m has a basis

(4.1.9b)
$$\mathfrak{M} = [\beta_1, \ldots, \beta_m]$$

chosen with coefficients a_{jk} from $\mathbf{Z}$ so that

(4.1.9c)
$$\begin{array}{l} \beta_1 = a_{11}\alpha_1 \\ \beta_2 = a_{21}\alpha_1 + a_{22}\alpha_2 \\ \quad \vdots \qquad \vdots \\ \beta_m = a_{m1}\alpha_1 + a_{m2}\alpha_2 + \cdots + a_{mm}\alpha_m \end{array}$$

with the inequalities

(4.1.9d)
$$0 \le a_{jk} < a_{kk}, \quad j > k.$$

Here β_1 is definable as one of the elements in the submodule $\mathbf{Q}\alpha_1 \cap \mathfrak{M}$ that minimizes the positive coefficients of α_1 to a_{11}; β_2 is likewise definable in the submodule $(\mathbf{Q}\alpha_1 + \mathbf{Q}\alpha_2) \cap \mathfrak{M}$ as minimizing (to a_{22}) the positive coefficients of a_2; and so on. The determinant in equations (4.1.9c) is half-triangular with a value representing the index of $\mathfrak{M}$ in O:

(4.1.9e)
$$j = [O : \mathfrak{M}] = a_{11} \ldots a_{mm}.$$

This would represent $N[\mathfrak{M}]$ if $\mathfrak{M}$ were an ideal.

4.1.10. Definition. *The "determinant" and the "discriminant" of a module $\mathfrak{M}$ are defined (respectively) as $D(\mathfrak{M})$ and $d(\mathfrak{M})$, where*

(4.1.10a)
$$D(\mathfrak{M}) = \pm \begin{vmatrix} \beta_1^{(1)} \cdots \beta_m^{(1)} \\ \cdot \\ \cdot \\ \cdot \\ \beta_1^{(m)} \cdots \beta_m^{(m)} \end{vmatrix}$$

so $D(\mathfrak{M}) = jD(O)$ (for positive j by choice of sign) and

(4.1.10b)
$$d(\mathfrak{M}) = D(\mathfrak{M})^2 = j^2 d(O).$$

Note that O is a special case of $\mathfrak{M}$ where matrix (4.1.9c) is the unit (diagonal ones). We call $d(O)$ the field-discriminant of K as contrasted with the root discriminant when $\mathfrak{M} = \mathbf{Z}[\theta]$. [Note that $D(\mathfrak{M})$ has an ambiguous sign because of the order in which the basis may be written, but $d(\mathfrak{M})$ does not.]

Thus, in searching for elements of O in $\mathbf{Z}[\theta]/|d(\theta)|$ [for equation 4.1.7], we need only test successive prime denominators such that $p^2 \mid d$. For convenience, consider how the module $\mathfrak{M}$ in equation (4.1.9b) of discriminant d would be reduced to O. We look for $\xi \in \mathfrak{M}/p$ such that $\xi \in O$ also. A typical such value would be

(4.1.10c)
$$\xi = (g_1\beta_1 + \cdots + g_r\beta_r)/p$$

where we can arrange to have $0 \leq g_i < p$ and $g_r = 1$ (by multiplying by a factor mod p). We then replace β_r by ξ and this produces a new module $\mathfrak{M}'$ $(\supset \mathfrak{M})$ whose discriminant is smaller by a factor of p^2. We continue to test a finite set of ξ until a minimum $|d(\mathfrak{M})|$ is reached. At this stage $\mathfrak{M}$ has grown to O.

4.1.11. Definition. *The "complex" imbedding of a field K of r real and $2s$ complex conjugates $(r + 2s = m)$ consists of the mapping from ξ (in K) into the $(r + s)$-tuple in m-dimensional space $S = \mathbf{R}^r \times \mathbf{C}^s$ of r real and s complex components,*

(4.1.11a)
$$\xi \to (\xi^{(1)}, \ldots, \xi^{(r)}, \xi^{(r+1)}, \ldots, \xi^{(r+s)}).$$

Clearly, the remaining s conjugates are merely complex conjugate values of the s complex components. When K is normal, then all fields are real $(r = m, s = 0)$ or all are complex $(r = 0, s = m/2)$. The integers form a discrete lattice in S (i.e., they consist of isolated points).

4.1.12. Definition. *The "real" imbedding of a field K of r real and 2s complex conjugate fields (m = r + 2s) consists of the mapping of ξ ($\in$ K) into the m-tuple in real space T = $\mathbf{R}^m$:*

(4.1.12a)

$$\xi \rightarrow (\xi^{(1)}, \ldots, \xi^{(r)}, Re\ \xi^{(r+1)}, Im\ \xi^{(r+1)}, \ldots, Re\ \xi^{(r+s)}, Im\ \xi^{(r+s)}).$$

As before, the integers are discrete in T and so is each submodule $\mathfrak{M}$. The basis of $\mathfrak{M}$ [see equation (4.1.9b)] consists of m vectors that form a fundamental parallelepiped whose volume is given by the determinant:

$$(4.1.12b) \quad \Delta(\mathfrak{M}) = abs \begin{vmatrix} \beta_1^{(1)}, \ldots, Re\ \beta_1^{(r+1)}, Im\ \beta_1^{(r+1)} \ldots \\ \cdot \qquad \cdot \qquad \cdot \\ \cdot \qquad \cdot \qquad \cdot \\ \cdot \qquad \cdot \qquad \cdot \\ \beta_m^{(1)}, \ldots, Re\ \beta_m^{(r+1)}, Im\ \beta_m^{(r+1)} \ldots \end{vmatrix}$$

$$(4.1.12c) \qquad\qquad \Delta(\mathfrak{M}) = |D(\mathfrak{M})/2^s|.$$

The advantage of a real imbedding is the use of geometry to show the existence of elements of any module of norm close to the origin.

4.1.13. Theorem (Minkowski). *Any module $\mathfrak{M}$ of (full) $\mathbf{Z}$-rank m contains a nonzero element ξ that lies within a rectangular parallelepiped,*

$$(4.1.13a) \quad \begin{cases} |\xi^{(t)}| \leq A_t, \quad t = 1, \ldots, r \\ |Re\ \xi^{(t+u)}| \leq A_{r+2u-1} \\ |Im\ \xi^{(t+u)}| \leq A_{r+2u}, \quad u = 1, \ldots, s \end{cases}$$

provided the volume V satisfies

$$(4.1.13b) \qquad\qquad V(= A_1 \ldots A_m) \geq \Delta(\mathfrak{M}).$$

In fact, the inequalities (4.1.13a) can be sharpened so that all but one are strict and (4.1.13b) also becomes strict.

4.1.14. Corollary. *Every module $\mathfrak{M}$ contains a nonzero element of norm less than $|D(\mathfrak{M})|$.*

4.1.14a. Corollary. *Every field discriminant (for K $\neq$ $\mathbf{Q}$) exceeds one.*

We omit the proofs, merely noting that these results are an effective measure of the discreteness of O. Likewise, we note the following theorem.

4.1.15. Theorem (Dedekind). *The ramified ideals of any field $K/\mathbf{Q}$ are precisely the divisors of the discriminant.*

This is a strengthening of the results deducible from Theorem 3.3.5, which allows for exceptional primes p (see Exercise 4.2). A major escalation of theory (ignored here) is required for Theorem 4.1.15. Incidentally, by virtue of Corollary 4.1.14a, there are always ramified primes in $K/\mathbf{Q}$ (unless $K = \mathbf{Q}$).

4.1.16. Theorem. *Every equivalence class of ideals has an ideal of norm less than $\sqrt{|d|}$, for d the field discriminant. Hence, the class number is finite.*

Proof. Consider an ideal class C and let $\mathfrak{a}$ belong to the inverse class C^{-1}. Then by Corollary 4.1.14, for some $(0 \neq) \xi \in \mathfrak{a}$, $|N\xi| < |D(\mathfrak{a})|$. But $(\xi) = \mathfrak{a}\mathfrak{b}$ for $\mathfrak{b}$ in C, and $|N(\xi)| = N[\mathfrak{a}]N[\mathfrak{b}] < |D(\mathfrak{a})| = |d|N[\mathfrak{a}]$. Thus, $N[\mathfrak{b}] < |d|$, and there is only a finite set of ideals $\mathfrak{b}$ that satisfies this description. (They are found from prime ideals of O that divide rational primes less than d.)   □

4.1.17. Remark. The metric estimates of discreteness contained in the last results are improvable; still they were not needed to prove the finiteness of the class number originally. Only Corollary 4.1.14a and Theorem 4.1.15 require strong estimates.

4.1.18. Remark. We wish to remain consistent with the terminology of quadratic field theory, where the discriminant d generates the field ($K = \mathbf{Q}(\sqrt{d})$). To do this we give the field discriminant the sign $(-1)^s$ of the root discriminant. Thus, in the quadratic case, $d > 0$ for the real field and $d < 0$ for the complex. In the cubic case, $d > 0$ for the totally real case ($r = 3$, $s = 0$) and $d < 0$ for the other case ($r = 1$, $s = 1$). For the biquadratic field, $d > 0$ in two cases, all real ($r = 4$, $s = 0$) and all complex ($r = 0$, $s = 2$), whereas $d < 0$ in the remaining case ($r = 2$, $s = 1$). Actually, for $m > 2$, the sign is of little value (since d is not a generating parameter of the field), so it is usually ignored.

The process of finding the class number so far seems to be to construct the ideal classes first (e.g., see Exercise 4.5). We introduce the notation $\mathrm{Cl}\{k\}$ for the class group and h for the class number so that

$$(4.1.19) \qquad\qquad h = |\mathrm{Cl}\{k\}|.$$

It will be seen later (Theorem 4.2.9) that h can be computed independently of the group structure. This requires other invariants in addition to the discriminant.

4.2. Units and regulators

The units O^* are a multiplicative subgroup of K. The torsion elements are roots of unity ($\rho^w = 1$) for some even w.

4.2.1. Illustration. For an imaginary quadratic field, the only units are roots of unity as follows:

$$d < -4, \quad O^* = \{\pm 1\} = \langle -1 \rangle, w = 2$$
$$d = -4, \quad O^* = \{\pm 1, \pm i\} = \langle i \rangle, \quad w = 4$$
$$d = -3, \quad O^* = \{\pm 1, (\pm 1 \pm \sqrt{-3})/2\} = \langle (1 + \sqrt{-3})/2 \rangle,$$
$$w = 6.$$

(We recognize the fourth and sixth roots of unity.)

4.2.2. Illustration. For a real quadratic field, the torsion units are always $\{\pm 1\}$ ($w = 2$), and a nontorsion unit exists, namely, ϵ. Thus,

$$O^* = \langle -1, \epsilon \rangle, \qquad \epsilon > 1$$

where $\epsilon = (t + u\sqrt{d})/2$ for $t > 0$, $u > 0$, and

$$(4.2.2a) \qquad t^2 - du^2 = \pm 4.$$

Here ϵ is the smallest value (> 1) that satisfies the so-called Pell equation (4.2.2a).

4.2.3. Theorem (Dirichlet). *The units O^* of a field K of degrees ($m = \rho$ $r + 2s$) are given by the generators*

$$O^* = \langle \epsilon_0, \epsilon_1, \ldots, \epsilon_g \rangle \quad (g = r + s - 1)$$

where ϵ_0 is the torsion unit of order w, and $\epsilon_1, \ldots, \epsilon_g$ are the so-called fundamental units that generate a free group. Its order g is called the (Dirichlet) torsion-free rank of the units. Any set of units of any rank is said to be "independent" if it generates a free group.

4.2.4. Definition. *The "logarithmic" imbedding of the nonzero elements of K (i.e., K) consists of the mapping of ξ ($\in K$) into $\Lambda(\xi)$ ($\in \mathbf{R}^{g+1}$); that is,*

$$(4.2.4a) \qquad \xi \to \Lambda(\xi) = (\lambda(\xi^{(1)}), \ldots, \lambda(\xi^{(r+s-1)}), \lambda(\xi^{(r+s)}))$$

where λ(ξ) is defined as

(4.2.4b)
$$\lambda(\xi^{(j)}) = log \ |\xi^{(j)}| \qquad (K^{(j)} \ real)$$
$$= 2 \ log \ |\xi^{(j)}| \qquad (K^{(j)} \ complex)$$

so that log $|N\xi| = \Sigma \ \lambda(\xi^{(j)})$ *summed over all g components. The "truncated" logarithmic imbedding is the corresponding map of* ξ *into* $\Lambda^*(\xi)$ $\in \mathbf{R}^g$ *formed by omitting the last vector component from* $\Lambda(\xi)$,

(4.2.4c)
$$\xi \rightarrow \Lambda^*(\xi) = (\lambda(\xi^{(1)}), \ldots, \lambda(\xi^{(r+s-1)})).$$

For U a set of g elements { . . . ξ_i . . . } *of K, the matrix whose rows are the g vector components of* $\Lambda^*(\xi_i)$ *and whose columns go from i = 1, . . . , g is called R(U), the "regulator" of U. In particular, if U represents a set of fundamental units, R(U) is called the "(field) regulator" R. The regulator also means the absolute value of the determinant of the said matrix. For g = 0 [e.g., see Illustration (4.2.1)], the null matrix has determinant 1, so by convention R = 1.*

It is easy to verify that a set of *g* units *U* has $R(U) \neq 0$ if and only if it is an independent set. If *U* is an independent set, *R(U)* is minimized by any fundamental set, so its value is invariant under the choice of the fundamental units. The regulator measures the fundamental volume for the logarithmic displacement of *O** under multiplication by the units, just as the determinant ($\sqrt{|d|}$) measures the fundamental domain under translation (see Definition 4.1.10). The discreteness of the logarithmic imbedding is expressed by the following theorem.

4.2.5. Theorem (Kronecker). *For each field there is a positive constant H such that for any integer* ξ, *each component of* $\Lambda(\xi)$ *is less than H in absolute value exactly when* ξ *is a torsion unit (*$\xi^w = 1$ *for some w). Indeed, the kernel of the mapping* $\xi \rightarrow \Lambda(\xi)$ *in O is the set of torsion units.*

This result follows from the discreteness of *O*; that is, if there are "too many" units close to one, then the units generate a set of integers that is "too crowded." The interesting contrast here is that if we have 𝔐, a submodule of *O*, and want to enlarge it to *O*, then the discriminant $|d(\mathfrak{M})|$ gives a practical working bound for trials (4.1.10c). Yet if we have an independent set of units *U* that we wish to enlarge to a fundamental set, then *R(U)* is of almost no help in practice. We are sometimes lucky to use units from smaller fields as an independent set (but there is no absolute rule, only a useful definition).

4.2.6. Definition. *Let a field K have the proper subfields k_t, each with its unit group O_t^*, and let U be the unit subgroup of O^* in K generated by these unit groups. Then $[O^*{:}U]$ is called the "unit (group) index" of K.*

If that index is finite, U is an independent set of units in K, but the construction of the full set O^* is not immediate. One frequent clue is that a unit of the subfield k_t that is totally positive might be a perfect square in K.

4.2.7. Remark. The knowledge of a set U of g independent units (not necessarily fundamental) and the knowledge of a full-rank module $\mathfrak{M}$ (not necessarily O) make possible the effective computation of the integers ξ of O whose absolute norm is less than some H. We immediately have a bound $d(\mathfrak{M})$ for denominators of $O \subseteq \mathfrak{M}/d$, so we look for elements of $\mathfrak{M}$ of norm $H|d|^m$. Then, using the units of U, we get a fundamental domain in the logarithmic imbedding in which all of $K^{\cdot}$ lies under translation by U. We look there among the discrete set of integers in $\mathfrak{M}$ of norm less than $H|d|^m$. At worst, there are superfluous solutions.

4.2.8. Remark. To pursue the earlier remark, if we can find all numbers of a given norm (to within associates), we can tell whether two ideals are equivalent. To test whether $\mathfrak{a} \sim \mathfrak{b}$, take $\mathfrak{b}'$ in the inverse class ($\mathfrak{b}\mathfrak{b}' \sim 1$) and test to see whether $\mathfrak{a}\mathfrak{b}'$ is principal. Find the minimal basis of $\mathfrak{a}\mathfrak{b}'$ so $N[\mathfrak{a}\mathfrak{b}'] = H$ is known. Then solve $N\xi = \pm H$ and see whether the ideal equality $\mathfrak{a}\mathfrak{b}' = (\xi)$ for any of the solutions by reducing each of the ideals to a common basis.

4.2.9. Density theorem (Dedekind). *Let $[K:\mathbf{Q}] = m \ (= r + 2s)$,*

$$d = \text{discriminant of } K$$
$$R = \text{regulator of } K$$
$$w = \text{number of roots of unity in } K$$
$$h = \text{class number of } K$$
$$\kappa = 2^{r+s}\pi^s R/(w\sqrt{|d|}) \quad \text{(Dirichlet's constant)}$$

Also let a be an ideal in O and define

(4.2.9a) $\qquad\qquad T_a(H) = Card\,\{(\xi) : |N\xi| \leq H,\ \xi \in a\}$

(4.2.9b) $\qquad\qquad T(H) = Card\,\{a : N[a] \leq H\}.$

Then

(4.2.9c) $\qquad\qquad\qquad lim_{H \to \infty} T_a(H)/H = \kappa/N[a]$

(4.2.9d) $\qquad\qquad\qquad lim_{H \to \infty} T(H)/H = \kappa h.$

This theorem is based on counting points ξ of O that lie in the ideal a but are reduced to avoid the repetition of associates. The details cannot be given here, but, in brief, the real imbedding makes the lattice points of a have a volume density of $1/(\sqrt{|d|}\ N[a])$, whereas the unit associates constitute a logarithmic imbedding of density $1/R$. The factors of π enter because the volumes are of circular regions, but the limit (4.2.9a, c) is a count of associates (ξ) by volume. The limit (4.2.9b, d) uses the fact that if $\xi \in a$, then $(\xi) = ab$ where $b \sim a^{-1}$, so that $T_a(H)$ counts the ideals in class $[b]$ with norm less than or equal to $H/N[a]$. Because $T(H)$ has all ideal classes, it obtains a factor of h.

4.2.10. Definition. *The "strict class number" h_+ is the order of the "strict class group" $Cl_+\{k\}$ based on a new equivalence, namely*

(4.2.10a) $\qquad\qquad\qquad\qquad a \approx b$

exactly when the quotient a/b is a principal ideal represented by a number with its conjugates positive for the r real conjugates of k. The "ordinary" class number (or group) is also called "weak."

For instance, in a quadratic field, if $d < 0$ (complex), the strict and weak classes are the same. If $d > 0$ (real), then $(\sqrt{d})$ and (1) are two ideals generated by elements with a negative norm $(-d)$ and a positive norm $(+1)$. These ideals are in the same strict equivalence class exactly when a unit of negative norm exists. (This happens when the Pellian equation has a solution with -4 in Illustration 4.2.2.)

4.2.11. Definition. *For every $\alpha \in K$, there is a "signature mapping" $\alpha \to (sgn\ \alpha^{(1)}, \ldots, sgn\ \alpha^{(r)})$ for the r real conjugates of α. This maps K into $C(2)^r$ (actually surjectively). If the mapping is restricted to O^*, the image is the subgroup $C(2)^{r-j}$. Here $r - j$ is the "signature rank" and j is the "signature defect."*

4.2.12. Theorem. *If j is the signature defect of the units of K, then*

$$(4.2.12a) \qquad h_+ = h2^j.$$

Also the strict class group has the (ordinary) weak class group as a subgroup with quotient group $C(2)^j$.

The proof follows from the fact that the signature mapping of K has defect 0. Furthermore, the squares of equivalent classes are strictly equivalent automatically. With the notation for class groups,

$$(4.2.12b) \qquad \mathrm{Cl}_+\{k\}/\mathrm{Cl}\{k\} = C(2)^j.$$

4.3. The zeta-function and *L*-functions

We now show how special Dirichlet series provide intuitions crucial to class field theory. These series are of the form $\sum_1^\infty a_n/n^s$ (where s no longer means the number of pairs of conjugate roots). The series is convergent in a half-plane, Re $s >$ const., but analytic details are omitted for brevity.

For the field K, the Dedekind zeta-function is

$$(4.3.1a) \quad \zeta(s, K) = \sum_{\mathfrak{a}} 1/(N\mathfrak{a})^s = \sum_{M=1}^\infty f(M)/M^s \qquad (\text{Re } s > 1)$$

where the first summation is over nonzero integral ideals $\mathfrak{a}$. Then

$$f(M) = \mathrm{Card}\,\{\mathfrak{a} : N\mathfrak{a} = M\}.$$

From equation (4.2.9d),

$$(4.3.1b) \qquad \lim_{M \to \infty} \left(\sum_1^M f(t) \right) \Big/ M = \kappa h.$$

A special case is the Riemann (and Euler) zeta-function (with $K = \mathbf{Q}$):

$$(4.3.1c) \qquad \zeta(s) = \sum_{M=1}^\infty 1/M^s \quad (\text{Re } s > 1).$$

Here $f(M) = 1$ ($\kappa = h = 1$). It can be verified that

$$(4.3.2a) \qquad \lim_{s \to 1} (s - 1)\zeta(s) = 1$$

and, analogously, by equation (4.3.2a), with Re $s > 1$ always,

$$(4.3.2b) \qquad \lim_{s \to 1} (s - 1)\zeta(s, K) = \kappa h.$$

To see equations (4.3.2a, b), define $S(M) = \Sigma_1^M f(t)$ so that

$$\zeta(s, K) = \Sigma\,(S(M) - S(M - 1))/M^s$$
$$= \Sigma\,(S(M)/M^s - S(M)/(M + 1)^s)$$
$$= \Sigma\,S(M)(1/M^s - 1/(M + 1)^s) = s\,\Sigma\,S(M)/M^{s+1} + \cdots$$

with negligible error as $s \to 1$. If $S(M)$ behaves like $M\kappa h$, then $\zeta(s, K)$ behaves like $\kappa h\zeta(s)$. On the other hand,

$$\zeta(s) = \int_1^\infty 1/[t]^s\,dt = \int_1^\infty 1/t^s\,dt + \cdots = 1/(s - 1) + \cdots$$

(with "lower-order error" as $s \to 1$ denoted by $\cdots$). Here $[t]$ denotes the highest integer that does not exceed t.

If we apply unique factorizations into primes in equations (4.3.1a, b), we obtain the Euler product decomposition:

$$(4.3.3)\qquad \zeta(s, K) = \Sigma\,1/(N\mathfrak{a})^s = \prod_\mathfrak{p}\,(1 + 1/(N\mathfrak{p})^s + 1/(N\mathfrak{p})^{2s} + \cdots)$$
$$= \Pi\,(1 - 1/(N\mathfrak{p})^s)^{-1}.$$

Taking logarithms, we find that

$$(4.3.4)\quad \log\zeta(s, K) = -\Sigma\,\log(1 - 1/(N\mathfrak{p})^s) = \Sigma\,1/(N\mathfrak{p})^s + \cdots.$$

Here we neglect error terms like $1/N(\mathfrak{p})^{2s}(1 - 1/(N\mathfrak{p})^s)^{-1}$, which converge for $s \geq 1$, by comparison with $\Sigma\,1/n^2$. Thus, by equations (4.3.2a, b),

$$(4.3.5)\quad \log 1/(s - 1) = \log\zeta(s, K) + \cdots = \log\zeta(s) + \cdots.$$

Therefore, if we define $H(p) = \mathrm{Card}\,\{\mathfrak{p} \mid \mathfrak{p}\colon N\mathfrak{p} = p\}$, then

$$(4.3.6)\qquad\qquad \Sigma\,H(p)/p^s + \cdots = \log 1/(s - 1).$$

For the Riemann zeta-function $H(p) = 1$,

$$(4.3.7)\qquad\qquad \Sigma\,1/p^s + \cdots = \log 1/(s - 1).$$

4.3.8. Definition. *For a set of primes P, the "(Dirichlet) density" is defined as*

$$(4.3.8a)\qquad lim_{s\to1}\left(\sum_{p\in P} 1/p^s\right)\Big/\left(\sum_p 1/p^s\right) = \delta(P).$$

[*This may in most cases equal the "linear" density, the limiting frequency ratio for primes in P, up to an increasing limit, but $\delta(P)$ is the only density that will be of interest here.*]

4.3.9. Theorem. *If $K/\mathbf{Q}$ is normal of degree m, then the splitting primes $P\,[= Spl(K)]$ are of density $1/m$.*

It follows from the fact that wfe (ramified primes) in equation (4.3.6) $H(p) = m$ if p splits and $H(p) = 0$ otherwise. The result shows that the set P cannot split in a normal field of higher degree. (This verifies Theorem 2.1.18.)

We next consider the effect of the ideal class group in the normal field $K/\mathbf{Q}$. Let J be the group of (fractional) ideals and let J^1 be the group of (fractional) principal ideals; then $H = J/J^1 = \mathrm{Cl}\{K\}$, the (finite) quotient group of order h, the class number. This group H has h characters $\chi_i(\mathfrak{a})$ defined over the ideals $\mathfrak{a}$, actually over the classes $[\mathfrak{a}]$. The characters have the following properties:

(4.3.10)
$$\left\{\begin{array}{l} \chi_i(\mathfrak{ab}) = \chi_i(\mathfrak{a})\chi_i(\mathfrak{b}) \\ \chi_i(\mathfrak{a})\chi_j(\mathfrak{a}) = \chi_k(\mathfrak{a}) \\ \chi_i(\mathfrak{a}) = 1 \text{ (for all } i) \Leftrightarrow \mathfrak{a} \sim 1 \\ \chi_i(\mathfrak{a}) = 1 \text{ (for all } \mathfrak{a}) \Leftrightarrow i = 1. \end{array}\right.$$

(Here the index k is uniquely determined by i and j and index 1 denotes the identity.) The characters take on values that are roots of unity of order h. Additionally, there are the following relations of orthogonality:

(4.3.11)
$$\left\{\begin{array}{l} \displaystyle\sum_{[\mathfrak{a}]}\chi_i([\mathfrak{a}])\chi_j^{-1}([\mathfrak{a}]) = \delta_{ij}h \\ \displaystyle\sum_{i}\chi_i([\mathfrak{a}])\chi_i([\mathfrak{b}])^{-1}) = \delta_{[\mathfrak{a}][\mathfrak{b}]}h. \end{array}\right.$$

Here the sums are over all ideal classes and all characters, while δ (the Kronecker delta) is one when the subscripts match and zero otherwise.

4.3.12. Definition. *The Hecke L-function with regard to a field K and an ideal group character χ is given by the sum over integral ideals of K (excluding 0 as usual)*

(4.3.12a) $\qquad L(s, K, \chi) = \displaystyle\sum_{\mathfrak{a}} \chi(\mathfrak{a})/(N\mathfrak{a})^s, \quad Re\,s > 1.$

Thus, $L(s, K, \chi_1) = \zeta(s, K)$.

If we follow the Euler product decomposition into primes p, then

(4.3.13a) $\quad L(s, K, \chi) = \displaystyle\prod_{\mathfrak{p}} (1 + \chi(\mathfrak{p})/(N\mathfrak{p})^s + \chi(\mathfrak{p})^2/(N\mathfrak{p})^{2s} + \cdots)$

and if we consider the limiting behavior as $s \to 1$ (as usual), then

$$(4.3.13b) \qquad \log L(s, K, \chi) = \sum_p \sum_{\mathfrak{p}|p} \chi(\mathfrak{p})/(N\mathfrak{p})^s + \cdots.$$

Note that the sum can again be limited to splitting primes ($N\mathfrak{p} = p$), since $\sum 1/p^2$ is bounded. For $\chi = \chi_1$, we again have equations (4.3.5) and (4.3.6), so

$$(4.3.14a) \qquad \log L(s, K, \chi_1) = \log 1/(s - 1) + \cdots.$$

It happens that for the other characters χ ($\neq \chi_1$), as $s \to 1$,

$$(4.3.14b) \qquad \log L(s, K, \chi) \to \text{limit} \ (\neq 0).$$

The convergence is not difficult, but the nonzero value of the limit is remarkably deep. We assume it all for now.

We now see, by orthogonality, that

$$(4.3.15) \qquad \sum_\chi \chi(\mathfrak{p}) = \sum_\chi \chi(\mathfrak{p})\chi(1) = h\delta_{\mathfrak{p}1}.$$

Hence, if we sum equation (4.3.13b) over all χ, we find from (4.3.14a, b) that

$$(4.3.16) \qquad \log 1/(s - 1) = h \sum I(p)/p^s + \cdots.$$

Here $I(p)$ counts the number of factors $\mathfrak{p}$ of degree 1 in p that are also principal. Since K is normal, if one factor is principal, so are the others (conjugates). Therefore, if P denotes $\mathrm{Spl}(K)$, the primes p that split into factors $\mathfrak{p}$ of degree 1, then $I(p)$ counts

$$(4.3.16a) \qquad P_1 = \{p : p \in P \text{ and factors of } p \text{ satisfy } \mathfrak{p} \sim 1\}.$$

Now $I(p) = m$ if $p \in P_1$ and 0 otherwise; so from equation (4.3.16),

$$(4.3.16b) \qquad \delta(P_1) = 1/mh.$$

From Theorem 4.3.9 we conclude the following;

4.3.17. Theorem. *Let K be normal over $\mathbf{Q}$ and let P be set $\mathrm{Spl}(K)$ so $K = CF\{P\}$. Let P_1 be the subset of P that splits into principal ideals. If $L = CF\{P_1\}$ exists, then L is of degree h relative to K; that is, $[L : K] = h$.*

The main result of this text (see Chapter 7.4) is that $(L =) CF\{P_1\}$ indeed exists. The analytic methods were necessary to place an upper bound on the degree of L (prior to any existence proofs). The "easier" result, which was superseded by equation (4.3.14b), was mere analyticity; that is,

$$(4.3.17a) \qquad L(s, K, \chi) = a(s - 1)^n + \cdots \qquad (\chi \neq \chi_1)$$

as $s \to 1$ for constant $a \neq 0$ and integer $n \geq 0$. If we check the steps, we see that with the result (4.3.17a), all we could conclude from Theorem 4.3.17 is that $[L : K] \geq h$. Remarkably, the (algebraic) information that $CF\{P_1\}$ exists is equivalent to the (analytic) information that (4.3.14b) is valid. In short (see Exercise 4.8), the connection is so tight that making the inequality $[L : K] \geq h$ into an equality is also equivalent to proving (4.3.14b). An illustration of the extended scope of the method of Hecke L-functions will follow.

4.3.17b. Remark. We have deliberately muted the role of the Galois theory, which will later become overwhelming. We might remark that the class field extension L/K has a Galois group isomorphic with the class group in K. Hence, class field extensions over K are abelian over K, and indeed such properties will later characterize abelian extension fields.

4.3.18. Remark on ring class fields. Let us start with the fundamental field discriminant d and consider primes p represented by the form $F_d(x, y)$ as in Chapter 2. Clearly, such primes have principal ideal factors in the integer ring O of $K = \mathbf{Q}(\sqrt{d})$ (see Remark 3.1.5). The converse requires a strict class structure (see Section 8.1).

4.3.18a. Theorem. *In the ring class field terminology (see Chapter 1),*

$$[RCF\{d\} : \mathbf{Q}(\sqrt{d})] = h_+.$$

Actually we have considered only field discriminants, but even for general discriminants df^2 there is an analogous result with h_+ interpreted in terms of a suitable class structure (see Section 8.1). We consider another instructive analog to a more general type of class structure.

4.3.19. Dirichlet's theorem on primes in arithmetic progression. *The initial work preceding the Hecke L-function was done by Dirichlet. Here χ is defined over $\mathbf{Z}$ on characters mod M (see Section 2.3) so $\chi(a) = 0$ if $(a, M) > 1$. We, of course, have a trivial "field extension" from $\mathbf{Q}$ to*

*K (= **Q**), but the concept of a principal ideal is in essence modified to be the fractional ideal for **Q** generated by numbers m (> 0) where m ≡ 1 mod M. There are h = φ(M) ideal classes and characters corresponding to (**Z**/M**Z**)* relatively prime residue classes mod M. We return in Section 7.2 to this (so-called ray) structure, but for now let us just note that for a prime p (> 0 always), all characters χ(p) = 1 exactly when p ≡ 1 mod M. Then*

(4.3.19a) $P_1 = \{p : p \equiv 1 \bmod M\}, \quad \delta(P_1) = 1/h.$

Because $\delta(P_1) > 0$, it follows that there are infinitely many primes in the arithmetic progression Mx + 1. By a variation of the argument, all h of the "coset" progressions Mx + a, (a, M) = 1, have infinitely many primes equidistributed with density 1/h.

*Thus, if the class field CF{P_1} exists, it will have to be of degree h over **Q**. Indeed, it is classically known to be **Q**(ρ), where ρ is the primitive Mth root of unity (exp 2πi/M) of degree h over **Q** (see Exercise 4.7).*

These analytic properties are of seemingly infinite variety and versatility and they will be briefly discussed again in Section 7.5. In summary, analytic methods establish an upper bound (hence uniqueness) for class fields and these methods provide a connection between two seemingly different objects: ordinary class structure and residue classes modulo *M*. It is the combination of these ideas that gives the main results in Chapter 7.

Exercises

4.1. Verify that for $f(\xi) = \xi^3 - a\xi^2 + b\xi - c = 0$, the root discrimination is $d = -27c^2 + a^2b^2 - 4b^3 - 4a^3c + 18abc$. *Hint:* First, d must be a linear combination of terms $a^r b^s c^t$, where $r, s, t \geq 0$ and $r + 2s + 3t = 6$. The coefficients of these terms are found by special cases. Thus, if $a = b = 0$, we can solve for the roots explicitly and see $d = -27c^3$; if $c = 0$, we solve a quadratic to find $d = a^2b^2 - 4b^3$; and finally, if $f(\xi) = (\xi - 1)^2(\xi \pm 1)$, then $d = 0$.

4.2. The field $K = \mathbf{Q}(\xi)$, where $\xi^3 - 2\xi^2 - 9\xi + 2 = 0$, has $d(\xi) = 62^2$, a perfect square, so Gal $K/\mathbf{Q}$ is normal [= $C(3)$]. Show that $(\xi^3 + \xi)/2$ is an integer (look at the equation) and $d = 31^2$. Because $N\xi = 2$, 2 has three different factors. From this, show that for all $\alpha \in O$, $j = [O : \mathbf{Z}[\alpha]]$ is even.

4.3. Show that for field discriminant d, the solutions t, u of the Pell equation (4.2.2a) produce an integer $\eta = (t + u\sqrt{D})/2$. Show that, to the contrary, if $t^2 - Du^2 = \pm 9$, then $(t + u\sqrt{D})/3$ would not necessarily be an integer. For the solutions to the Pell equation, verify that $|\eta|$ is minimized (for all $|\eta| > 1$) by the fundamental unit. Likewise verify that (among possibly several) with the minimum t (> 0), the fundamental unit minimizes t among $t > 0$ (but not always uniquely). Note that $t = (\eta + \eta')/2$. Show that if $d \equiv 0 \bmod 8$, then $t \equiv u \equiv 0 \bmod 2$.

4.4. Let Gal $K = D(n)$, the dihedral group; then $[K : Q] = |D(n)| = 2n$, where $D(n) = \langle S, T : S^n = T^2 = 1, TS = S^{-1}T\rangle$. Thus, for $n = 2$, $D(2)$ is abelian $[= C(2) \times C(2)]$. Then K has the subfield k_0 of order 2 (fixed by the cyclic group $\langle S\rangle$ of order n) and the subfields k_j ($j = 1, \ldots, n$) of order n (fixed by the group $\langle ST^j\rangle$ of order 2). We take n prime so that these are all the subgroups and subfields. Then this identity is valid for all A ($\neq 0$) in K:

$$A^n = (AA^T)\cdots(AA^{TS^{n-1}})/(A^T\cdots A^{TS^{n-1}}) = \alpha_1\ldots\alpha_n/\alpha_0$$

where $\alpha_j \in k_j$. From this show that the group index is either 1 or n. For $n = 2$ and $K = Q(\sqrt{N}, \sqrt{M})$, we can choose k_j as $Q(\sqrt{N})$, $Q(\sqrt{M})$, and $Q(\sqrt{MN})$, respectively, for $j = 0, 1$, and 2. Therefore for any unit η in K, $\eta^2 = \epsilon_0\epsilon_1\epsilon_2$ for units ϵ_j in k_j. Find cases among small negative N and M such that the unit index is 1 or 2 as desired.

4.5. Show that the class number of $Q(\sqrt{79})$ is 3 by testing the classes of divisors of primes less than $\sqrt{d} = 2\sqrt{79}$ (see Theorem 4.1.16). Of course, 11 and 17 are inert, $(79/17) = -1$, so we test $p = 2, 3, 5, 7, 13$. Since $2 = 2_1^2$ and $2_1 = (9 + \sqrt{79})$, $\epsilon = \frac{1}{2}(\sqrt{79} + 9)^2 = 80 + 9\sqrt{79}$ (unit). If we set $f(x) = x^2 - 79 = N(x - \sqrt{79})$, we see that the primes factor according to the values $f(1) = -78, f(2) = -75, f(3) = -70, f(4) = -63$, and $f(5) = -54$. Thus, $(1 - \sqrt{79}) = 2_1 3_1 13_1$, $(2 - \sqrt{79}) = 3_2 5_1^2$, $(3 - \sqrt{79}) = 2_1 5_2 7_1$, $(4 - \sqrt{79}) = 3_1^2 7_2$, and $(5 - \sqrt{79}) = 2_1 3_2^3$. Note that the subscripts (in the factorization $p = p_1 p_2$) identify homomorphisms as follows: $2_1 \Rightarrow \sqrt{79} \equiv 1 \bmod 2_1$, $3_1 \Rightarrow \sqrt{79} \equiv 1 \bmod 3_1$, but $3_2 \Rightarrow \sqrt{79} \equiv 2 \bmod 3_2$, and so on. Then, from $2_1 \sim 1$, it follows from $f(5)$ that $3_1^3 \sim 1$, and from the other $f(n)$ that factors 5, 7, and 13 are in the class group generated by 3_1. All that remains is to show that 3_1 is not principal; that is, that $\pm 3 = x^2 - 79y^2$ has no solutions in $\mathbf{Z}$. Check for solutions where $x, y > 0$ and $|\theta| = |x + y\sqrt{79}|$ lies between 1 and $\sqrt{\epsilon}$. (Replace θ by $\theta\epsilon^t$ for a suitable t.) At the same time verify that ϵ is fundamental by looking for solutions of $\pm 1 = x^2 - 79y^2$ subject to the same range. (It pays to restrict the search by taking the norm equations modulo 3 and 79.)

4.6. Show that for field discriminant d, prime p is represented by $F_d(x, y)$ (as in Chapter 2) precisely when p splits into two principal ideals in $\mathbf{Q}(\sqrt{d})$, assuming $d < 0$, or $d > 0$ and $N\eta = -1$ for the unit η. [Of course, assume $(p, 2d) = 1$.]

4.7. Show that for $K = \mathbf{Q}(\rho)$, where $\rho = \exp 2\pi i/M$, p splits completely from $\mathbf{Q}$ to K [for $(p, M) = 1$] exactly when $p \equiv 1 \bmod M$. [Note that Gal $K/\mathbf{Q}$ consists of automorphisms $\rho \to \rho^t$ for $(t, M) = 1$, so Gal $K/\mathbf{Q} = (\mathbf{Z}/M\mathbf{Z})^*$, an abelian group of order $h = \phi(M)$.] Let $\mathfrak{P} \mid p$ and $N[\mathfrak{P}] = p^f = |O/\mathfrak{P}|$ (the order of the finite field). We have to ask when $f = 1$, or $\xi^p \equiv \xi \bmod \mathfrak{P}$ for all ξ in O. Actually, it suffices to take $\xi = \rho$, so we need only use the fact that $\rho^t \equiv \rho \bmod \mathfrak{P} \leftrightarrow t - 1 \equiv 0 \bmod M$.

The same argument applies if $K = k(\rho)$, where k need not be "prime" to $\mathbf{Q}(\rho)$, that is, possibly $k \cap \mathbf{Q}(\rho) \neq \mathbf{Q}$. In any case, K/k is normal. Let $\mathfrak{p} \mid p$ in k ($p \nmid M$) and make the further assumption that $\mathfrak{p}$ is of degree 1; that is, $N\mathfrak{p} = p$. Then any α in O_k satisfies $\alpha \equiv a \bmod \mathfrak{p}$ for $a \in \mathbf{Z}$, and consequently any $A \in O_K$ satisfies $A \equiv \xi \bmod \mathfrak{P}$ for ξ in $\mathbf{Q}(\rho)$, where $\mathfrak{P}$ (in K) divides $\mathfrak{p}$. Show that $\mathfrak{p}$ splits completely from k to K exactly when $(p =) N\mathfrak{p} \equiv 1 \bmod M$.

4.8. Suppose, for all $\chi \neq \chi_1$, we knew only equation (4.3.17a) rather than the stronger result (4.3.14b). Show that we could then have only one χ for which $n > 0$ (and for this χ $n = 1$), and, if so, $\delta(P_1) = 0$ and $CF\{P_1\}$ does not exist. (See Theorem 4.3.9.) Proceeding in greater ignorance, knowing only that $L(s, k, \chi)$ remains finite for all $\chi \neq \chi_1$, draw the conclusion that $[L : K] \geq h$.

4.9. The number of lattice points inside the circle $x^2 + y^2 \leq H$ is approximately πH (the area). Show that this verifies the density theorem (4.2.9a) when $K = \mathbf{Q}(i)$.

4.10. Prove equation (4.2.9b) directly from (4.2.9a) by showing as an intermediate step that if, for ideal class C,

$$T_C(H) = \text{Card } \{\mathfrak{b} : \mathfrak{b} \in C, \quad N[\mathfrak{b}] \leq H\}$$

then $T_C(H)/H \to \kappa$. [Note that $(\xi) = \mathfrak{a}\mathfrak{b}$ in equation (4.2.9a).]

4.11. Show by the argument of equation (4.3.16) and the related equations that there are infinitely many primes $\mathfrak{p}$ of K (of degree 1) that are in a given ideal class determined (say) by $\mathfrak{a}$. Indeed, instead of summing equation (4.3.13b) over χ, we sum $\chi(\mathfrak{a}^{-1}) \log L(S, K, \chi)$. Thus, we obtain

$$\log 1/(s - 1) = \Sigma \chi(\mathfrak{a}^{-1}) \log L(s, K, \chi) + \cdots$$
$$= h \sum_{\mathfrak{p} \epsilon P} I_{\mathfrak{a}}(p)/p^s + \cdots$$

where $I_{\mathfrak{a}}(p) = \text{Card } \{\mathfrak{p} \,|\, p : \mathfrak{p} \in P, \mathfrak{p} \sim \mathfrak{a}\}$ so $I_1(p) = I(p)$ ($= 0$ or m), but, more generally, $0 \leq I_{\mathfrak{a}}(p) \leq m$. What is the density statement analogous with equation (4.3.16b)? (Compare Theorem 7.5.5 in Chapter 7.)

Bibliographic note

The best references are the standard introductory texts or monographs, for example, Weber (1899), Landau (1918), Hecke (1923), Herbrand (1936), Weil (1967), Lang (1970), Gundlach (1972), and Riebenboim (1972). The change of basis problem in operations (3.1.8) is discussed in Jacobson (1974). For geometric results, see Minkowski (1896); also compare Tate (1950). A sampling of the use of the zeta-function in computing class numbers is seen in Heilbronn (1967), Masley (1977), and Meyer (1957).

5

Function fields

The algebraic numbers needed for class field theory are generated by modular functions that arise by a process not yet completely explained (even by the proofs). Modular functions come only at the second level of conceptualization, with algebraic functions coming first. It is easy to see at the first level that algebraic functions, defined generally over **C**, lead to "new" algebraic numbers when the coefficients and the independent variables are restricted to "old" algebraic numbers. The procedure is more complicated, however. On a more abstract level we have "equivalence classes" of algebraic functions, and these classes are characterized by modular functions. It is these modular functions (belonging to "classes" of elliptic functions) that generate our required algebraic numbers.

Returning to more general theory, we make the observation that Dedekind rings arose as a generalization of integers in two types of fields:

1. algebraic number fields, and
2. algebraic function fields of one independent variable.

The selection process of the integer ring for number fields makes possible the unique factorization into ideals. When the process is extended to functions fields, it becomes simpler because power series are the basic tool. Going back to number fields, we adapt the use of power series in the form of "local theory," which is used advantageously later on.

5.1. Rational functions

We consider the algebraic function field over $\mathbf{C}(z)$,

$$(5.1.1a) \qquad\qquad \mathcal{F} = \mathbf{C}(z, w)$$

which is of degree of transcendence 1. Thus, one variable, say z, may be viewed as "independent" and the other defined by an irreducible polynomial over $C(z)$ with rational functions as coefficients, namely,

$$(5.1.1b) \qquad w^n + A_1(z)w^{n-1} + \cdots + A_n(z) = 0.$$

The field $\mathcal{F}$ consists of functions

$$(5.1.1c) \qquad \mathcal{F} = \{f(z, w): f \in C(z, w)\}.$$

In fact, $f(z, w)$ can be restricted to polynomials of degree less than n in w with rational functions of z as coefficients.

The preceding definition seems to depend on the variables that generate $\mathcal{F}$. To look at $\mathcal{F}$ invariantly, we say that another field $C(Z, W)$, where W and Z have a relation like equation (5.1.1b), is (birationally) "equivalent" to $C(z, w)$ when every element of one is in the other. This means that

$$(5.1.2) \qquad \begin{cases} z = g(Z, W), & w = h(Z, W) \\ Z = G(z, w), & W = H(z, w) \end{cases}$$

for g, h, G, H all rational functions [in $C(z, w)$ or $C(Z, W)$].

A function field $\mathcal{F}$ is called "rational" if

$$(5.1.3) \qquad \mathcal{F} = C(z).$$

[For consistency, we could think of $\mathcal{F} = C(z, w)$ with $w = z$.] Thus, the field

$$(5.1.4a) \qquad \mathcal{F} = C(z, w), \qquad w^2 + z^2 = 1$$

is rational $[= C(t)]$ by a classical parametrization (of "Pythagoras"),

$$(5.1.4b) \qquad \begin{cases} z = 2t/(1 + t^2) \\ w = (1 - t^2)/(1 + t^2) \\ (1 + t)/(1 - t) = (1 + z)/w. \end{cases}$$

Of course, the last equation easily expresses t in z and w. More generally,

$$(5.1.5) \qquad C(t) = C((at + b)/(ct + d))$$

for a, b, c, d in C with $ad - bc \neq 0$. Of all the rational functions, only the linear fractional transformations are reversible. So the rational function fields can be expressed in terms of a so-called uniformizing parameter determined with equation (5.1.5), and all such fields are birationally equivalent. (No modular functions occur yet!)

The functions $f(z)$ in $\mathbf{C}(z)$ are characterized by the fact that they are "meromorphic" at all z, including $z = \infty$. This means that if we substitute $z = 1/Z$, $f(z)$ becomes meromorphic in Z at $Z = 0$. Otherwise expressed, the variable Z can be defined as a "local" variable for any complex (finite) value a like $Z = (z - a)$. Then, in terms of the suitable local variable, if z is in a neighborhood of each finite or infinite point.

$$(5.1.6) \qquad |f(z)| < C|Z|^{-T}$$

for positive constants C and T (varying from point to point). This property characterizes rational functions $f(z)$ as functions with poles as the only singularities [i.e., expression (5.1.6)]. The domain of definition that consists of finite points and ∞ is called the "z-sphere" or the "Riemann sphere."

Here we must recall that the definition of "local" variables enables us to define a (general) "analytic manifold $\mathcal{M}$" as the collection of (infinitely many) points $\{P, Q, \ldots\}$ and neighborhoods of each point, $\{N(P), N(Q), \ldots\}$ with variables $Z_P, Z_Q, \ldots$ defined in each. These are called "local" (unformizing) parameters. They have the following consistency property: If $N(P)$ and $N(Q)$ intersect, then in the intersection Z_P and Z_Q will be related by analytic functions in a one-to-one fashion:

$$(5.1.7a) \qquad Z_P = G(Z_Q), \qquad Z_Q = H(Z_P).$$

Then a function f is rational if it has "elements" $f_P(Z_P)$ and $f_Q(Z_Q)$ that are meromorphic [i.e., that satisfy boundedness conditions (5.1.6)] for each neighborhood in such a way that the definitions are consistent. This means that

$$(5.1.7b) \qquad f_P(Z_P) = f_Q(Z_Q)$$

when Z_P and Z_Q are related by equations (5.1.7a). Of course, we do not use infinitely many local variables in practice. A finite number of regions is thought of as the set of neighborhoods of the manifold with relations valid at the common overlap (triangulation). The set of regions is called an "atlas." A compact manifold $\mathcal{M}$ is one that has no boundary.

We can now define $\mathcal{M}$ and $\mathcal{M}'$ as "equivalent" manifolds if the points P of $\mathcal{M}$ and P' of $\mathcal{M}'$ with neighborhoods $N(P)$ and $N(P')$ can be biuniquely mapped onto each other in a manner consistent with the overlap relations (5.1.7a) on $\mathcal{M}$ and $\mathcal{M}'$. Thus, with $Z_P \in N(P)$ in $\mathcal{M}$ and $Z_{P'} \in N(P')$ in $\mathcal{M}'$, there are the mappings

$$(5.1.7c) \qquad Z_{P'} = f(Z_P), \qquad Z_P = g(Z_{P'}).$$

Note that a function "defined" on $\mathcal{M}$ will mean just a rational (or meromorphic) function.

If we refer again to the z-sphere as a manifold, it may be described in terms of an atlas of two neighborhoods of 0 and ∞, namely,

(5.1.8a) $Z_0 = z$ for $|z| < 2$, $Z_\infty = 1/z$ for $|z| > \frac{1}{2}$.

This provides an intersecting region $\frac{1}{2} < |z| < 2$. A function defined on the sphere consists of two elements $f_0(Z_0)$ and $f_\infty(Z_\infty)$ related in this intersection by

(5.1.8b) $f_0(Z_0) = f_\infty(Z_\infty)$, $Z_0 = 1/Z_\infty$.

In this way both elements are defined over a finite domain, although they may have infinite values. Incidentally, if we were to omit the point at ∞ (or any other points) from the definition, we would have a noncompact manifold. In practice the only examples that occur in this text are compact or easily compactifiable by adjoining an obvious "missing" point. We assume compactness.

It is important to note that for the mainfold $\mathcal{M}$, there is a well-defined order of zero of pole for each function, regardless of which element is taken for the definition at the point. It is the nature of a one-to-one mapping that the derivative is not zero.

We now return to the z-sphere to illustrate definitions that can be made with general manifolds in mind. Every nonzero function that is meromorphic on the z-sphere is representable (everywhere but at ∞) by its zeros a_i and its poles b_j as

(5.1.9a) $$f(z) = \text{const.} \frac{(z - a_1)^{u_1} \cdots (z - a_r)^{u_r}}{(z - b_1)^{v_1} \cdots (z - b_s)^{v_s}}.$$

We view this locally by treating all points alike and by treating zeros and poles alike in a common notation. We write $D(a)$ to be the "divisor" at a in a purely formal sense so that a zero of order u is $uD(a)$ and a pole of order v is $-vD(a)$ (with u or v possibly zero, of course, when either zero or pole fails to occur). In the preceding example, if U and V are the total orders of zeros and poles as shown in equation (5.1.9a) so that

$$U = \Sigma\, u_i, \qquad V = \Sigma\, v_j$$

we can write the zeros and poles in equation (5.1.9b) in a formal manner. We must also consider that if $Z = 1/z$ (the variable at ∞), then f behaves like Z^{V-U}, written $(V - U)D(\infty)$ to cover a zero $(V > U)$ or a pole $(U > V)$ or a regular point $(U = V)$. Symbolically, $f(z) = f$ is written as the divisor

(5.1.9b) $(f) = \Sigma u_i D(a_i) - v_j D(a_j) + (V - U)D(\infty)$.

For a nonzero constant f, the coefficient at each point is zero and so is the divisor.

5.1.10. Definition. *For a given manifold $\mathcal{M}$, the "divisors" are the set of finite formal sums $\mathcal{D} = \Sigma\, uD(a)$ over points a of $\mathcal{M}$ with $u \in Z$. Each u is the "order" at a. The sum of the orders $\Sigma\, u$ is the "degree." We accordingly write $\mathrm{ord}_a\mathcal{D}$ for each u and $\deg \mathcal{D}$ for the degree.*

Each nonzero function f that is rational (meromorphic) on $\mathcal{M}$ then leads to a unique divisor $\mathcal{D}$ corresponding to its zeros and poles. Thus, if $(f_1) = \mathcal{D}_1$ and $(f_2) = \mathcal{D}_2$, then

(5.1.11) $(f_1 f_2) = \mathcal{D}_1 + \mathcal{D}_2$, $(f_1/f_2) = \mathcal{D}_1 - \mathcal{D}_2$.

Of course, $(k) = 0$ for a nonzero constant k. As before, we write $\mathrm{ord}_a(f)$ and $\deg(f)$ for the order of f at a and degree of the divisor (f). From the earlier expression (5.1.9a,b), we conclude Theorem 5.1.12.

5.1.12. Theorem. *For a z-sphere a function f is generally definable by any divisor of degree 0, and f is unique to within a nonzero constant factor.*

5.1.13. Definition. *For any manifold $\mathcal{M}$, a divisor $\mathcal{D}$ is "positive" if ord $\mathcal{D} \geq 0$ at every point of $\mathcal{M}$. It is written $\mathcal{D} > 0$. These divisors form a semigroup. (We also write $\mathcal{D} < 0$ if $-\mathcal{D} > 0$.)*

5.1.14. Definition. *A "linear space" of $\mathcal{D}$, written $L(\mathcal{D})$, is the space of all rational functions on a manifold $\mathcal{M}$ that have at most the poles indicated by positive orders and at least the zeros indicated by the negative orders in $\mathcal{D}$; that is,*

$$L(\mathcal{D}) = \{f : (f) + \mathcal{D} > 0\}.$$

The "dimension," written dim $\mathcal{D}$, is the dimension of the vector space over $\mathbf{C}$ formed by $L(\mathcal{D})$. As a consequence of the definition, for any $f (\neq 0)$,

(5.1.14a) $\dim \mathcal{D} = \dim((f) + \mathcal{D})$.

As a further consequence, if $\mathcal{D} > 0$, then dim $\mathcal{D} \geq 1$, since the complex constants are then always present in $L(\mathcal{D})$.

For an arbitrary compact $\mathcal{M}$, not the z-sphere, Theorem 5.1.12 has only a weak analog for now.

5.1.15. Theorem. *If f is meromorphic on $\mathcal{M}$, then (f) is a divisor of degree 0.*

The proof involves the technique of triangulation. If $\mathcal{M}$ is described in terms of a finite set of neighborhoods overlapping in triangular (or polygonal) boundaries (chosen to avoid zeros and poles), then within each triangle T of $\mathcal{M}$, the zeros and poles form a divisor $\mathcal{D}_T$ whose degree is given by Cauchy's residue theorem as

$$(5.1.15a) \qquad \deg \mathcal{D}_T = \int_T [f'(z)/2\pi i f(z)] \, dz.$$

Now (f) is the sum of all $\mathcal{D}_T$ and $\mathrm{ord}(f) = 0$ because the integrals over adjoining triangles cancel (from the compactness of $\mathcal{M}$).

5.1.16. Theorem. *For the z-sphere, dim $\mathcal{D} = \deg \mathcal{D} + 1$ if deg $\mathcal{D} \geq 0$, and dim $\mathcal{D} = 0$ if deg $\mathcal{D} < 0$.*

For proof observe that if $\mathcal{D}$ is positive, $\mathcal{D} = u_i D(a_i) + u_\infty D(\infty)$, where $u_i \geq 0$ and $u_\infty \geq 0$. Then $L(\mathcal{D})$ is the set

$$(5.1.16a) \qquad L(\mathcal{D}) = \{p(z)/\Pi \, (z - a_i)^{u_i}\}$$

where $p(z)$ is a polynomial of $\mathbf{C}[z]$ of degree less than or equal to deg $\mathcal{D}$. Such $p(z)$ constitute a space of dimension deg $\mathcal{D} + 1$ [the number of coefficients of $p(z)$]. If $\mathcal{D}$ is not positive, we write it as a difference:

$$(5.1.16b) \qquad \mathcal{D} = \mathcal{D}_1 - \mathcal{D}_2, \qquad \mathcal{D}_1 > 0, \quad \mathcal{D}_2 > 0.$$

When deg $\mathcal{D} > 0$, we can find an f such that $(f) + \mathcal{D} > 0$ by taking $(f) = \mathcal{D}_0 - \mathcal{D}_2$, where $\mathcal{D}_0$ consists of a subset of terms of $\mathcal{D}_1$. This is possible because of Theorem 5.1.12. Then note $\deg((f) + \mathcal{D}) = \deg \mathcal{D}$, and $\dim((f) + \mathcal{D}) = \dim \mathcal{D}$. When deg $\mathcal{D} < 0$, similarly, we can replace $\mathcal{D}$ by $(f) + \mathcal{D}$, which is a negative divisor, so $L(\mathcal{D})$ consists of only 0 and dim $\mathcal{D} = 0$.

It is a crucial result that for an arbitrary $\mathcal{M}$, this theorem is replaced by the more difficult Riemann–Roch theorem (in Section 5.2).

5.2. Riemann surfaces

The "Riemann surface" $\mathcal{M}$ is an analytic manifold constructed with reference to the field $\mathcal{F} = C(z, w)$ of equations (5.1.1a, b). Because w is of degree n over $C(z)$, we construct an n-sheeted covering of the z-sphere on which w can be defined uniquely, and so can each element of $\mathcal{F}$. Although the manifold $\mathcal{M}$ is constructed with special reference to z and w, the aggregate of points is a birational invariant of $\mathcal{F}$. [This is true in the same way that the field $k = Q(\alpha) = Q(\beta)$ is a set of numbers defined independently of whether k is generated by α or β.]

In general, the defining relation for w is a polynomial in $C[z, w]$, assumed irreducible:

(5.2.1a) $E(z, w) = a_0(z)w^n + a_1(z)w^{n-1} + \cdots + a_n(z)w^0 = 0.$

Here "in general" n values of w are defined for each z. They are distinct except possibly for a finite number of values, namely, the simultaneous roots of

(5.2.1b) $E(z, w) = \partial E(z, w)/\partial w = 0$

and also possibly roots of $a_0(z) = 0$ and $z = \infty$ (which we treat by changing variables to $z = 1/Z$). If these points are exlcuded, we can subdivide the z-sphere into a finite number of triangles in which there are n distinct solution functions $w_i(z)$. Each triangle is considered to be one of n replicas, on each of which $w_i(z)$ is single-valued.

Under analytic continuation of a circle about the vertex α in the triangulation, however, the $w_i(z)$ values are permuted and the permutation can be broken down into several cycles. Each of the cycles about $z = \alpha$ is regarded as a point of the Riemann surface over $z = \alpha$. Suppose $n = 6$ and the $w_i(z)$ values ($i = 1, \ldots, 6$) permute like

(5.2.2)   (123)(56): $w_1 \rightarrow w_2 \rightarrow w_3 \rightarrow w_1;$ $w_4 \rightarrow w_4;$ $w_5 \rightarrow w_6 \rightarrow w_5.$

Then a change of variables $z - \alpha = Z^3$ will make the cycle (123) single-valued in Z. Simultaneously, the change of variables $z - \alpha = Z'^2$ makes the cycle (56) single-valued in Z'. At the same time, w_4 is single-valued. Thus, there are three (not six) values of w over $z = \alpha$, or three (not six) points of the Reimann surface. The degree of ramification for each point

(5.2.2a) $z - \alpha = Z^e$ (or $z = 1/Z^e$ for $\alpha = \infty$)

is $e - 1$ (which vanishes at an ordinary point). This is the number of points of the Riemann surface that are "lost" by each branching. The total ramification of $\mathcal{M}$ is

(5.2.2b) $$b = \Sigma\,(e - 1).$$

Because ∞ is also covered, $\mathcal{M}$ is compact. Also $\mathcal{M}$ is connected; that is, each of the n values of w must be obtainable from any other by continuation because of the irreducibility of the defining equation $E(z, w)$.

5.2.3. Theorem. *A function f defined on a Riemann surface $\mathcal{M}$ takes each value the same number of times.*

Proof. Consider for α finite, the divisor

$$(f - \alpha) = \mathcal{D}_+ - \mathcal{D}_-$$

where $\mathcal{D}_+$ and $\mathcal{D}_-$ are both positive divisors that stand for the zeros and poles, respectively. Now $\mathcal{D}_+$ depends on α, but $\mathcal{D}_-$ does not, since $f - \alpha = \infty$ at the same points regardless of α. If $N = \deg \mathcal{D}_-$, the number of poles, then so does $N = \deg \mathcal{D}_+$, regardless of α. $\qquad\square$

5.2.3.a. Corollary. *A bounded function defined on a Riemann surface $\mathcal{M}$ is a constant.*

5.2.4. Remark. It is important to note that the Riemann surface can, in retrospect, be constructed by a triangulation of the z-sphere into an n-fold covering, by geometrical fiat. We need not know any function $w(z)$ algebraic over z of degree n. We can still speak as before of divisors $\mathcal{D} = \Sigma\, mD(a)$ without necessarily having functions to represent the divisors (see Remark 5.2.7). It is an important result that any such construction of a manifold $\mathcal{M}$ is still supported by algebraic functions $w(z)$ (see Corollary 5.2.9). This idea is very important in modular function theory, where the Riemann surface is usually constructed as a "geometric house" for which functions will be found that "live inside."

The manifold $\mathcal{M}$ does not determine the representation $\mathcal{F} = \mathbf{C}(z, w)$ because $\mathcal{F} = \mathbf{C}(Z, W)$ for other pairs of equivalent variables. Various numbers, such as the sheets n, the branchings e, and the ramification b, depend on the defining variable. The most important invariant of $\mathcal{M}$ is the (unique) topological invariant, the "genus" g. It is defined in two significantly related ways.

1. The genus g is the number of "handles" that $\mathcal{M}$ has when represented as a nonintersection surface, by reassembling the triangles (which are somehow "entwined" over the z-sphere). An equivalent definition is that it takes $2g$ closed curves (the "cross-cuts") to cut up $\mathcal{M}$ into a simply connected region. In terms of any specific covering,

$$(5.2.5a) \qquad g = -n + 1 + b/2.$$

In principle, this is seen from the fact that every pair of branch points makes for a handle if we do not count the $(n - 1)$ pairs that are needed simply to connect the n sheets. In this way the cross-cuts $\mathcal{C}_1, \ldots, \mathcal{C}_{2g}$ form a homology basis for all closed curves on $\mathcal{M}$ (with coefficients in $\mathbf{Z}$). Thus, if $\mathcal{C}$ is any closed curve, then

$$(5.2.5b) \qquad \mathcal{C} \sim n_1 \mathcal{C}_1 + \cdots + n_{2g} \mathcal{C}_{2g}, \qquad n_j \in \mathbf{Z}.$$

2. The second definition of genus is the number of abelian integrals (of the first kind) independent over $\mathbf{C}$ not counting constants of integration. These integrals are defined as expressions $\int f \, dz$, with f defined on $\mathcal{M}$, that are always finite but (necessarily) multivalued (see Corollary 5.2.3a). They are a vector space over $\mathbf{C}$ that can be written in terms of a $\mathbf{C}$-basis as

$$(5.2.5c) \quad f(z, w) \, dz = \alpha_1 f_1(z, w) \, dz + \cdots + \alpha_g f_g(z, w) \qquad (\alpha_i \in \mathbf{C}).$$

The differential is useful because, in terms of the local variable Z at P, $f(z, w) \, dz = F(Z) \, dZ$, so the test is whether $F(Z)$ is finite (i.e., analytic) at each $P(Z = 0)$. Combining equations (5.2.5a) and (5.2.5b), we have

$$(5.2.5d) \qquad \int_{\mathcal{C}} f(z, w) \, dz = \sum_{ij} A_{ij} \alpha_i n_j$$

where the A_{ij} are the so-called Riemann periods:

$$(5.2.5e) \qquad A_{ij} = \int_{\mathcal{C}_j} f_i(z, w) \, dz.$$

If we standardize the g abelian differentials, we have a vector of $2g$ complex numbers for its integrals over $2g$ homology basis paths. These $2g \times g$ periods [see equation (5.2.5e)] form the so-called Riemann matrix.

5.2.6. Illustration. We can achieve any genus by the equation

$$(5.2.6a) \qquad w^2 = (z - r_1) \cdots (z - r_m)$$

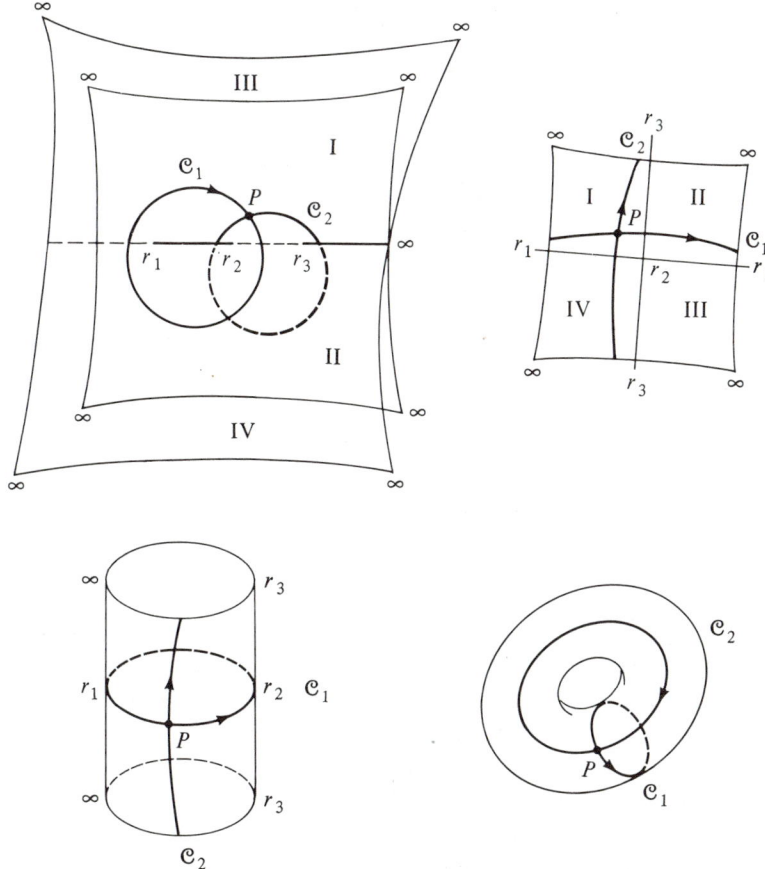

Figure 5.1. Riemann surface for $w^2 = (z - r_1)(z - r_2)(z - r_3)$ over the z-sphere, showing topological equivalence with torus (genus 1).

for m distinct roots r. Now $n = 2$ and $b = m$ for m even and $b = m + 1$ for m odd (where ∞ is the other branch point). Thus (see Figure 5.1 for $m = 3$),

$$(5.2.6b) \qquad g = \begin{cases} (m - 2)/2, & m \text{ even} \\ (m - 1)/2, & m \text{ odd.} \end{cases}$$

The case $g = 0$ is the z-sphere [equivalent to $\mathbf{C}(t)$]; the case $g = 1$ is the torus or period parallelogram (elliptic function field, see Section 5.3); and the case $g > 1$ is called "hyperelliptic." [It is worth remarking that equation (5.2.6a) gives the most general field $\mathbf{C}(z, w)$ for g up to

2, but not for $g > 2$.] The homology basis is shown in the figure and the abelian integrals are

$$(5.2.6c) \quad \int df(z, w) = \int z^t \, dz/w, \qquad t = 0, 1, \ldots, g - 1.$$

As an illustration of the divisor notation, note that

$$(5.2.6d) \quad (w) = D(r_1) + \cdots + D(r_m)$$
$$- (m/2)D(\infty_1) - (m/2)D(\infty_2)$$

for m even (with two points of M over ∞ as shown), and

$$(5.2.6e) \quad (w) = D(r_1) + \cdots + D(r_m) - mD(\infty)$$

for m odd (with only one point of M over ∞). Likewise, note that

$$(5.2.6f) \quad (z - r_1) = 2D(r_1) - D(\infty_1) - D(\infty_2), \qquad m \text{ even}$$
$$(5.2.6g) \quad (z - r_1) = 2D(r_1) - 2D(\infty), \qquad m \text{ odd}.$$

5.2.7. Remark. Equation (5.2.6g) shows that $(z - r_1)^{1/2}$ seems to be given by the divisor $\mathcal{D} = D(r_1) - D(\infty)$ for m odd, a divisor with deg $\mathcal{D} = 0$. Nevertheless, no function of $\mathbf{C}(z, w)$ satisfies $(f) = \mathcal{D}$ when $g > 0$. This will be proved in Section 5.3, but in the meantime, a positive existence result can be given, which is a theorem of profound interest in many branches of mathematics.

5.2.8. Riemann–Roch theorem. *If $\mathcal{M}$ is a compact Riemann surface of genus g, then for any divisor $\mathcal{D}$ on $\mathcal{M}$,*

$$dim \ \mathcal{D} \geq deg \ \mathcal{D} + 1 - g.$$

This is the existence theorem par excellence. For instance, consider some consequences:

5.2.9. Corollary. *Let a compact Riemann surface $\mathcal{M}$ be constructed not from a relation like $E(z, w) = 0$ [of equation (5.2.1a)] but geometrically, out of a covering of the z-plane by triangles that fit together to form a manifold. Then there exists an algebraic function w of z such that the relation $E(z, w) = 0$ [or the field $\mathcal{F} = \mathbf{C}(z, w)$] generates $\mathcal{M}$.*

5.2.10. Corollary. *Let Z be any nonconstant function of the field $\mathcal{F} = \mathbf{C}(z, w)$; then there exists a function W, such that $\mathcal{F} = \mathbf{C}(Z, W)$.*

Thus (see Theorem 5.1.16 for genus 0), if $\mathcal{D} > 0$ and deg $\mathcal{D} \geq g + 1$, then dim $\mathcal{D} \geq 2$, or a nonconstant w exists that has no poles other than those accounted for in $\mathcal{D}$ [i.e., $(w) + \mathcal{D} > 0$]. If we set all these $(g + 1)$ poles to occur at $(g + 1)$ points of $\mathcal{M}$, all lying over different points of the z-sphere, then some w exists with $0 < N \leq (g + 1)$ poles only at the subset of these points. We assert now that $C(z, w)$ generates $\mathcal{M}$ (see Corollary 5.2.9). To prove Corollary 5.2.10, we similarly assert that W should be defined as having its poles at points where the variable Z has different values.

To complete both these proofs, we must consider one additional concept.

5.2.11. Definition. *A function w defined on a Riemann surface $\mathcal{M}$ over the z-sphere is called "primitive" (with respect to z) when w generates $\mathcal{M}$. Then a general point of $\mathcal{M}$ can be written as (z, w), meaning that z is given and the branch of w over z is given.*

If W is not primitive on an n-sheeted surface $\mathcal{M}$, then only $N (<n)$ conjugates of W occur, and on analytic continuation, they replicate themselves n/N times (indeed n must have N as a factor). In proving Corollary 5.2.9, note that the value $w = \infty$ does not occur more than once over each z, so w is primitive. (The same holds for Corollary 5.2.10.)

5.3. Elliptic functions

We consider the field of genus 1, $\mathcal{F} = C(z, w)$, over $\mathcal{M}$,

$$(5.3.1) \qquad w^2 = 4z^3 - g_2 z - g_3 = 4(z - r_1)(z - r_2)(z - r_3)$$

where r_1, r_2, and r_3 are distinct (finite) points. The discriminant is given (compare Exercise 4.1) by

$$(5.3.2a) \qquad (r_1 - r_2)^2 (r_2 - r_3)^2 (r_3 - r_1)^2 = (g_2^3 - 27 g_3^2)/16.$$

Thus, we are assuming that

$$(5.3.2b) \qquad g_2^3 - 27 g_3^2 \neq 0.$$

This is later seen to be the most general Riemann surface of genus 1. It is topologically a torus or a sphere with one handle. It is also topologically a parallelogram with opposite sides identified, that is, a doubly periodic structure. All such structures are topologically equivalent but not analytically equivalent, as we shall see.

The abelian integral (of the first kind) is

$$(5.3.3) \qquad\qquad u = \int_{\infty}^{z} dz/w$$

and leads to a system generated by two periods: $\int_{\mathcal{C}_1}$ and $\int_{\mathcal{C}_2}$ (see Figure 5.1). Thus,

$$2\omega_1 = \int_{\mathcal{C}_1} = 2\int_{r_1}^{r_2} dz/w$$

$$2\omega_2 = \int_{\mathcal{C}_2} = 2\int_{r_2}^{r_3} dz/w.$$

As we note, w becomes $-w$ on the segment $r_1 r_2$ or $r_2 r_3$ when it is approached from the other side by the circuit $\mathcal{C}_1$ or $\mathcal{C}_2$. If we formally define a lattice $\mathcal{L}$ of periods by assuming Im $\omega_1/\omega_2 \neq 0$, it can be written as

$$(5.3.4a) \qquad \mathcal{L} = [2\omega_1, 2\omega_2] = \{\Omega\}$$
$$\Omega = 2n_1\omega_1 + 2n_2\omega_2 \qquad (n_1, n_2 \in \mathbf{Z})$$
$$(5.3.4b) \qquad \tau = \omega_1/\omega_2 \qquad (\text{Im } \tau > 0).$$

Thus, we can say that for $\Omega \in \mathcal{L}$,

$$(5.3.4c) \qquad\qquad \Omega \equiv 0 \bmod \mathcal{L}.$$

The system of periods $\mathcal{L}$ defines a parallelogram II that covers the u-plane with a doubly periodic structure. This manifold II is the one constructed for w and z, but that does not prove that w and z are definable biuniquely on it. (Indeed, w and z were originally defined on the Riemann surface over the z-sphere.) We redefine w and z on II. Let

$$(5.3.5) \qquad p(u) = 1/u^2 + \sum_{\Omega}{}' [1/(u - \Omega)^2 - 1/\Omega^2]$$

the (Weierstrass) p-function, as long as the series converges uniformly away from the poles (at Ω). (Here Σ' means a sum with the $1/0$ term omitted.) We can prove the following equation for $p(u)$:

$$(5.3.6a) \qquad p'(u)^2 = 4p(u)^3 - g_2 p(u) - g_3$$
$$(5.3.6b) \qquad g_2 = 60 \ \Sigma' \ 1/\Omega^4$$
$$g_3 = 140 \ \Sigma' 1/\Omega^6$$
$$\Omega = 2n_1\omega_1 + 2n_2\omega_2.$$

If we expand both sides of equation (5.3.6a) about $u = 0$, which (by translation) is the only singularity, we see that the difference vanishes there, whereas it is analytic everywhere else. So, by Corollary 5.2.3a, both sides agree.

Hence, in equation (5.3.1), if g_2 and g_3 can be expressed in the form of equations (5.3.6b), then by solving simultaneously for suitable periods $2\omega_1$ and $2\omega_2$, it follows that $z \, [= p(u)]$ and $w \, [= p'(u)]$ are definable from the polynomial (5.3.1) as well as the periods $\mathcal{L}$. Also $p(u) = \infty$ twice on $\mathcal{M}$ (at $u = 0$), so $p(u)$ takes every value twice, as does z (recall that the Riemann surface $\mathcal{M}$ is two-sheeted over z). The two values of z belong to $z = p(u) = p(-u)$. Because $w = p'(u) = -p'(-u)$, the choice of u versus $-u$ (or w versus $-w$) makes the pair (z, w) correspond to a unique value of u in Π. (See Exercise 5.4.)

There are two existence theorems for functions on Π. The first is (Riemann–Roch) Theorem 5.2.8 for genus 1:

5.3.7. Theorem. *If deg $\mathcal{D} \geq 1$, then dim $\mathcal{D} = deg \, \mathcal{D}$.*

For $\mathcal{D} > 0$, this means that as long as two or more poles are prescribed (possibly with multiple poles), then a nonconstant function f exists with at most these poles; we can even say, with exactly these poles. Actually, a more exact assignment of poles and zeros can be made:

5.3.8. Theorem (Abel). *The divisor $\mathcal{D} = \Sigma \, e_i D(u_i)$ belongs to an elliptic function f [i.e., $(f) = \mathcal{D}$) if and only if*
1. deg $\mathcal{D} = 0$ $(\Sigma \, e_i = 0)$
2. $\Sigma \, e_i u_i \equiv 0 \bmod \mathcal{L}$.

The necessity of this theorem is proved by taking $\int d(\log f)/2\pi i$ about the boundary of Π for condition 1 and $\int u \, d(\log f)/2\pi i$ for condition 2. The existence part of both theorems can be established by an explicit construction. We restrict the proofs, however, to $\mathcal{D} > 0$ (for Theorem 5.3.7), so we need to consider only an assignment of poles. Weierstrass constructed two functions, not quite doubly periodic:

(5.3.9) $\zeta(u) = -\int p(u) \, du = 1/u + \Sigma' \, (1/(u - \Omega) + 1/\Omega + u/\Omega^2)$
(5.3.10) $\sigma(u) = \exp \int z(u) \, du = u\Pi'(u - \Omega) \exp(u/\Omega + u^2/2\Omega^2).$

These functions have an additive and multiplicative constant (respectively) under the period operation; that is,

(5.3.9a) $\zeta(u + \Omega) = \zeta(u) + \eta(\Omega), \qquad \eta(\Omega)$ constant
(5.3.10a) $\sigma(u + \Omega) = \sigma(u)\gamma(\Omega), \qquad \gamma(\Omega)$ constant.

So, in accordance with Riemann–Roch, if $\mathcal{D} > 0$ and all poles are simple so that $\mathcal{D} = \Sigma\, D(u_i)$, then Theorem 5.3.7 follows from

(5.3.11) $L(\mathcal{D}) = \{a_0 + \Sigma\, c_i\zeta(u - u_i): \Sigma\, c_i = 0\}.$

(The case of multiple roots is given as Exercise 5.5.] For Abel's theorem,

(5.3.12) $f(u) = \text{const } \Pi_i\sigma(u - u_i)^{a_i}.$

5.3.13. Theorem. *Every field $\mathcal{F}$ of genus 1 can be generated by a cubic polynomial [like equation (5.3.1)].*

The proof is a consequence of Riemann–Roch (Theorem 5.3.7). A nonconstant function (say z) exists with two assigned poles. Then we find another (say w, by Corollary 5.2.10) such that $C(z, w) = \mathcal{F}$. Now w must have two sheets over the z-sphere because z is two-valued. From Illustration 5.2.6, we see that a cubic or biquadratic polynomial must be used. (See Exercise 5.3.)

In other words, any field $\mathcal{F}$ defined over a parallelogram (corresponding to a lattice $\mathcal{L}$) will come from a cubic polynomial whose lattice of periods is $\mathcal{L}$.

We are dealing with so-called modular relationships, of establishing "modules" to represent the equivalence classes of fields of genus 1. If we start with the polynomials that define a field of genus 1, we have two possible forms:

(5.3.14a) $w^2 = (z - r_1)(z - r_2)(z - r_3)(z - r_0)$
(5.3.14b) $w^2 = (z - r_1)(z - r_2)(z - r_3) = z^3 - az^2 + bz - c$

where the roots are unequal and finite [in equation (5.3.14b), by analogy, $r_0 = \infty$]. For either form only the cross-ratio λ is invariant under linear fractional transformations (preserving the order of the roots), where

(5.3.15a) $\lambda = (r_0 - r_1)(r_2 - r_3)/(r_0 - r_3)(r_2 - r_1)$

or

(5.3.15b) $\lambda = (r_2 - r_3)/(r_2 - r_1) \qquad (r_0 = \infty).$

Considering the 4! orders in which the roots could be numbered, the four substitutions $\{1, (10)(23), (12)(03), (13)(02)\}$ preserve each λ. Therefore, λ takes six values under the renumbering of the roots:

(5.3.16) $\{\lambda_i\} = \{\lambda, 1/\lambda, 1 - \lambda, 1/(1 - \lambda), \lambda/(\lambda - 1), (\lambda - 1)/\lambda\}$.

We form a symmetric function of these λ as an invariant. First of all, $\Sigma \lambda_i = 0$. So the simplest to define is the "Klein" invariant, using $\Sigma \lambda_i^2$:

(5.3.17a)
$$J[\lambda] = 2/27\Sigma \lambda_i^2 + 6/27$$
$$= 4(\lambda^2 - \lambda + 1)^3/27(\lambda - 1)^2\lambda^2.$$

The constants are chosen for the later convenience (of having desirable multiple factors). Likewise, for number-theoretical convenience, we shall choose the "Weber" invariant (later on), defined by

(5.3.17b) $j = 1728J$.

The cross-ratio λ is a function of the roots [see equation (5.3.15b)] and so is $J[\lambda]$, but $J[\lambda]$ is symmetric in λ_i and hence in r_i. Indeed, after calculation [referring to equation (5.3.14b)],

$$J[\lambda] = 4(a^2 - 3b^3)^3/27d$$

for d is the discriminant (see Exercise 4.1). Thus, in reference to the cubic in the form of equation (5.3.1),

(5.3.18) $J = g_2^3/(g_2^3 - 27g_3^2)$

so J is evidently a finite complex number for each cubic equation.

When g_2 and g_3 come from a period lattice [see equations (5.3.6b)], it is not immediately clear that the denominator in equation (5.3.18) is not zero. (This is shown in Chapter 10, where an explicit expression is found for it.) Thus, J is a function of $\mathcal{L}$ (= $\{\Omega\}$), the aggregate of periods, so we can write

(5.3.19) $J = J[\lambda] = J\{\Omega\}$.

In terms of the basis of $\mathcal{L}$, there are equivalences created by

(5.3.20a) $\omega_1 \to \omega_2 + \omega_1, \quad \omega_2 \to \omega_2$
(5.3.20b) $\omega_1 \to \omega_2, \quad \omega_2 \to -\omega_1$.

All unimodular changes of basis are generated by the earlier equivalences (see Chapter 10 for further details). Indeed, g_2 and g_3 remain unchanged by these transformations because they constitute a re-arrangement of terms in a sum. Also, if $\lambda \in \mathbf{C}$ ($\lambda \neq 0$), then

$$(5.3.19a) \qquad J = J\{\Omega\} = J\{\lambda\Omega\}.$$

We can finally write

$$(5.3.19b) \qquad J = J(\tau), \qquad \tau = \omega_1/\omega_2, \qquad \mathrm{Im}\ \tau > 0$$

because from equation (5.3.18), J is homogeneous in ω_1, ω_2 of degree 0, as the ratio of two polynomials in g_1, g_2. Thus, we are saying that the "size" of the period parallelogram Π does not matter; only the "shape" does. Indeed, many configurations form equivalent shapes if we put expressions (5.3.20a, b) in terms of τ:

$$(5.3.21a) \qquad \qquad \tau \to \tau + 1$$
$$(5.3.21b) \qquad \qquad \tau \to -1/\tau.$$

These transformations generate the "modular group" that preserves the upper half τ-plane [hence the minus sign in expression (5.3.21b)]. In terms of J,

$$(5.3.21c) \qquad J(\tau) = J(\tau + 1) = J(-1/\tau).$$

The exact makeup of the modular group is deferred to Chapter 10.

5.3.22. Theorem. *Any two lattices $\mathcal{L}$, $\mathcal{L}'$ are equivalent ($\mathcal{L} = \lambda\mathcal{L}'$, $\lambda \neq 0$) exactly when the corresponding basis forms ratios τ, τ' connected by the modular group.*

5.3.22a. Corollary. *Two elliptic function fields are birationally equivalent exactly when the period parallelograms of the abelian integrals have basis ratios equivalent under the modular group.*

The corollary requires that the abelian integrals λu are an invariant of the field (to within the proportionality factor λ). Conversely, the abelian integral defines the same field through the $p(u)$ function. We still need an additional fact: In Chapter 10, it will be proved (see Theorem 10.3.5) that $J(\tau)$ takes on each (finite) complex number once for each equivalence class of τ under the modular group. Therefore, $J(\tau)$ [or the related $j(\tau)$ in equation (5.3.17b)] serves as the much promised modular invariant of the (elliptic function) fields of genus 1.

5.3.23. Illustration. There are two special symmetric cases:

$$(5.3.23a) \qquad w^2 = 4z^3 - 4 \qquad (J = 0, j = 0)$$
$$(5.3.23b) \qquad w^2 = 4z^3 - 4z \qquad (J = 1, j = 1728).$$

In the first case, the period structure is invariant under rotation by ρ = $\exp 2\pi i/3$, since z appears only as a cube. We can multiply the z-sphere by ρ without changing w, so $du = dz/w$ is multiplied by ρ. Thus, $g_2 = \rho^{-4}g_2 = 0$. In the second case, if we multiply z by -1, we multiply w by i, so the period structure is invariant under rotation by i and, similarly, $g_3 = i^{-6}g_3 = 0$.

5.4. Integer rings

It is valuable to draw an analogy between algebraic number field extensions $K/\mathbf{Q}$ and algebraic function field extensions $\mathbf{C}(z, w)/\mathbf{C}(z)$. The number-theoretic extensions are richer in invariants, but the invariants are more complicated. The idea of the analogy is to simplify the viewpoint for number fields.

For instance, given $K/\mathbf{Q}$ an extension of degree $n = [K:\mathbf{Q}]$, we see that the role of $\mathbf{Q}$ is absolute; it is the minimal field in K. On the other hand, given $\mathcal{F} = \mathbf{C}(z, w)$, there is really no way to unequivocally extract $\mathbf{C}(z)$ by some minimal property of $\mathcal{F}$. The best we might do is minimize n, the number of sheets of the Riemann surface of w over z. The corresponding invariant is like the genus g. By the Riemann–Roch theorem, $n \leq g + 1$, and as we shall see, $1 + g$ is the analog of class number. Therefore, in an intuitive way, algebraic function fields are a model of extensions whose degree is a class number, that is, of class fields (see Section 4.3).

Let us begin the analogy with the concept of "integer." For a rational field $\mathcal{F} = \mathbf{C}(z)$, there is the obvious choice of $\mathbf{C}[z]$, the polynomials. More abstractly, if $\mathcal{M}$ denotes the z-sphere and $\mathcal{M}_P$ denotes $\mathcal{M}$ with P removed (so $\mathcal{M}_\infty$, as a set, is $\mathbf{C}$), then the integer ring is

$$(5.4.1) \qquad O_\infty = \{f : f \in \mathcal{F}, \quad (f) > 0 \text{ on } \mathcal{M}_\infty\} (= \mathbf{C}[z]).$$

Clearly this choice of integer ring is not invariant in an absolute sense, since $\mathbf{C}(z) = \mathbf{C}(t)$ if $t = (az + b)/(cz + d)$, so $z = \infty$ can go into any value of t. We could just as well define integers O_P as the set where $(f) > 0$ on $\mathcal{M}_P$.

In general, if $\mathcal{M}$ is any (compact) algebraic Riemann surface, then we let S be an arbitrary finite so-called singular set, and let $\mathcal{M}_S$ be the manifold $\mathcal{M}$ "punctured" by the removal of S.

5.4.2. Definition. *The set O_S of "S-integers" on the Riemann surface $\mathcal{M}$, corresponding to a function field $\mathcal{F}$, is given by*

$$(5.4.2a) \qquad O_S = \{f : f \in \mathcal{F}, \quad (f) > 0 \text{ on } \mathcal{M}_S\}.$$

If $\mathcal{F} = \mathbf{C}(z, w)$ and w is of degree n over $\mathbf{C}(z)$, then z is covered by a Riemann surface of n sheets, so in a very "natural" way there are n (or fewer) points over $z = \infty$ to serve as S. Then the S-integers are the elements of $\mathcal{F}$ that satisfy monic equations with coefficients in $\mathbf{C}[z]$. This is far from invariant, since an equivalent form of $\mathcal{F} = \mathbf{C}(Z, W)$ could have the points over $z = \infty$ transformed into points scattered all over the Riemann surface of W over the Z-sphere. It is much more natural to find a representation $\mathcal{F} = \mathbf{C}(z, w)$ where there is just one point over $z = \infty$ (branched so $e = n$). Thus, $S = \{\infty\}$ is a single point.

5.4.3. Theorem. *Any algebraic function field $\mathcal{F}$ defined on $\mathcal{M}$ may be transformed birationally to the form $\mathbf{C}(z, w)$ where any prescribed point P of $\mathcal{M}$ lies (alone) over $z = \infty$ as a branch point of order equal to the number of sheets $n \{= [\mathbf{C}(z, w):\mathbf{C}(z)]\}$.*

This is a result of the Riemann–Roch theorem, with $\mathcal{D} = (g + 1) D(P)$ (compare Theorem 5.2.8). Then if we define O_∞, it is a ring of integers in $\mathcal{F}$ containing $\mathbf{C}[z]$. By general results in Section 3.2, O_∞ is a Dedekind ring, with unique factorization into prime (maximal) ideals. A more elaborate examination of the ring reveals the following theorem.

5.4.4. Theorem. *The ideals in O_∞ correspond to positive divisors in $\mathcal{M}_\infty$. Thus, the ideal I corresponds uniquely to the divisor $\mathcal{D} = \Sigma\, a_i D(P_i)$ for P_i the set of (finite) points at which all of the functions in the ideal I must vanish.*

5.4.5. Corollary. *The prime (or equivalently, the maximal) ideals P of O_∞ correspond uniquely to finite points P of the Riemann surface with the property that $f \in P$ if and only if $f(P) = 0$. In fact, if the point P lies above $(z =) \alpha$ on the z-sphere, then $P = (z - \alpha, f_P)$, in the two-element form of the ideal P, for a suitable f_P.*

At this point the so-called local theory becomes very powerful. The function f_P in the corollary is characterized by the fact that it vanished with a zero of first order in Z_P, the local parameter at P. In fact, if $Z_P^e = z - \alpha$, then the ideal $(z - \alpha)$ factors as ΠP^e where P goes over ideals for all points that lie over α. This is a simplistic approach because it uses power series in Z, a variable not generally in F. Yet this is what we shall do in Chapter 6 to study the general branching of extension number fields.

5.4.6. Theorem. *For a field of genus 1, the ideal classes of O_∞ correspond uniquely to the points of the Riemann surface.*

For proof, take the points of the period parallelogram $\mathcal{M}$, so $u = 0$ corresponds to $P = \infty$. Then consider $\mathcal{D} = \Sigma\, a_i D(P_i)$ on $\mathcal{M}_\infty$ with $u = u_i$ corresponding to the point P_i of $\mathcal{M}_\infty$. Then, by Abel's theorem, $\mathcal{D} = (f)$ for some function f on $\mathcal{M}$ exactly when $\Sigma\, a_i u_i \equiv 0$ mod $\mathcal{L}$. Thus, the values of u (mod $\mathcal{L}$) in the period parallelogram constitute a module for equivalence classes of $\mathcal{D}$. All values of u are involved because any $D(P)$ is a legitimate divisor on $\mathcal{M}_\infty$.

5.4.7. Remark. The same result holds for any genus g. There are infinitely many ideal classes of complex dimension g. This also follows from a generalized form of Abel's theorem. When $g = 0$, of course, $\mathcal{F}$ is a rational field with class number one.

To achieve a finite class number, it is necessary to use as coefficient field not $\mathbf{C}$, but a finite field. We do not do this here.

Exercises

5.1. In reference to Remark 5.2.7, show that in $\mathbf{C}(z, w)$, where $w^2 = (z - a)(z - b)(z - c)$, no W can exist that satisfies $W^2 = z - a$.

5.2. Find the parametrization to permit $\mathbf{C}(z, w) = \mathbf{C}(t)$, where $w^2 = (z - r_1)(z - r_2)$.

5.3. Find the change of variables to transform $\mathbf{C}(z, w)$ to $\mathbf{C}(Z, W)$, where $w^2 = (z^2 - 1)(k^2 z^2 - 1)$ and $W^2 = Z(Z - 1)(Z - \lambda)$. What is the (cross-ratio) relation between k and λ?

5.4. The half-periods ω of the parallelogram are ω_1, ω_2, and $\omega_1 + \omega_2$. These correspond to values of $p(u)$ at the roots r_i of the cubic in equation (5.3.1). Show that the divisors $\mathcal{D} = D(\omega) - D(0)$ for each of the half-periods can satisfy $2\mathcal{D} = (f)$ but not $\mathcal{D} = (f)$. (The divisors $\mathcal{D}$

correspond to the classical elliptical functions sn u, cn u, and dn u, which are single-valued only in a *double* parallelogram.)

5.5. Give an explicit formula like equation (5.3.11) for $L(\mathcal{D})$ when $\mathcal{D}$ has multiple divisors. Note that for multiple values of a, $\zeta(u - a)$ involves successive derivatives with respect to a, that is, $p(u - a)$, $p'(u - a)$, and so on.

5.6. Define the invariant polynomial $\Pi_i(X - \lambda_i) = X^6 - A_1 X^5 + A_2 X^4 - \cdots + A_6$. Clearly, $A_1 = 3$ and $A_6 = 1$, but what are the other coefficients in terms of J?

Bibliographic note

Algebraic functions are discussed in various related contexts in Fricke and Klein (1890, vol. 2), Fricke (1928), and Van der Waerden (1950); also see Artin and Tate (1952) and Swinnerton-Dyer (1974). The analytic and topological preliminaries are found in Ahlfors and Sario (1960). For further classical reference to elliptic functions and fields, see Whittaker and Watson (1940).

6

Relative fields

We are now prepared to generalize the concept of splitting a prime in $K/\mathbf{Q}$ to the given field extension K/k of degree n. We say that the prime ideal $\mathfrak{p}$ (in k) "splits" in K if $\mathfrak{p}$ factors in K as the product of n distinct primes $\mathfrak{P}$ (in K). Moreover, if $\mathfrak{p}$ does not split from k to K, it is desirable to know for which intermediate fields L between k and K the ideal $\mathfrak{p}$ will split. This quest leads to a succession of "mileposts" between k and K, indicating the splitting status of some factor $\mathfrak{P}$ of $\mathfrak{p}$. We refer to the Hilbert sequence in Section 6.2.

The splitting property is very deep and is connected with group characters such as the quadratic residue symbol. No greater evidence of the depth of this property need be given than a gratuitous proof of quadratic reciprocity in Section 6.3.

Meanwhile, we take note of a portentous feature of the splitting concept. The ramified primes in K/k are excluded from consideration, and they seem unimportant because they are finite in number. They do not, for example, affect "density" theorems (as in Section 4.3) or rules of decomposition of primes (which are stated with a finite list of exceptions). We cannot escape from ramified primes, however, because such special primes help characterize the "general" primes that split from k to K. (Examples are given in Section 6.4.)

Therefore, we must track down the primes that are ramified in K/k and even find them in the analogous "relative" discriminant of this extension. To do so becomes all the more difficult because there is no general way of writing $O = o[A]$ as a monogenic ring, where O is the integer ring of K and o is that of k. Worse still, there is no minimal basis for the ring O over the ring o in general (despite the fact that, as we saw in Section 4.1, each of these rings has a minimal basis over $\mathbf{Z}$).

What we do is to simulate the desired monogenic behavior by arguing that $O = o[A]$ is still true in a "local" sense, and this enables us to study the ramified primes (see Section 6.2). Again, we remark that this approach is inspired by the facility of "local" parameters in function theory.

6.1. Discriminant and different

Consider any extension field $K \supset k$ (normal or not) with $[K : k] = n$. If $A \in K$ and $A^{(1)} (= A)$, $A^{(2)}$, ..., $A^{(n)}$ are conjugates, these lie in the normal (Galois) closure $\hat{K}$ ($= K$ for K normal). The integer rings are O_K and O_k. The concepts of trace and norm generalize for K/k as follows:

(6.1.1a) $$Tr_{K/k}A = \sum_i A^{(i)}$$

(6.1.1b) $$N_{K/k}A = \prod_i A^{(i)}.$$

Thus, we generate $K = k(A)$ using integers for convenience:

(6.1.2) $$f(A) = A^n + a_1A^{n-1} + \cdots + a_n (= 0), \qquad a_i \in O_k.$$

This leads to a new definition of relative (root) different $\delta_{K/k}$ and also the (root) discriminant $d_{K/k}$ consistent with Section 4.1:

(6.1.3a) $$\delta_{K/k}(A) = \prod_{j=2}^{n} (A^{(1)} - A^{(j)}) = f'(A) \quad (\in K)$$

(6.1.3b) $$d_{K/k}(A) = \prod_{i>j} (A^{(i)} - A^{(j)})^2 = \pm N_{K/k}f'(A) \quad (\in k).$$

We can also introduce the relative (root) determinant $D_{K/k}$ as

(6.1.4a) $$d_{K/k}(A) = D_{K/k}(A)^2$$

(6.1.4b) $$D_{K/k}(A) = \det(a_{ij}), \qquad a_{ij} = A^{(i)j-1}.$$

By the Vandermonde theorem, we have (see Section 4.1):

(6.1.4c) $$d_{K/k}(A) = \det(b_{ij}), \qquad b_{ij} = Tr_{K/k}A^{i+j-2}.$$

If $k = \mathbf{Q}$, these concepts coincide with the unsubscripted "absolute" (root) different, discriminant, or determinant.

The different (which is a derivative in principle) is multiplicative under the chain rule. For instance, if (say)

$$(6.1.5) \quad \begin{array}{c} K = L(A) \supset L = k(B) \supset k, \\ [K:L] = N, \quad [L:k] = m, \quad [K:k] = Nm\,(=n) \end{array}$$

and, furthermore, if A and B can be defined as monic equations:

$$(6.1.6a) \quad f(B) = B^m + b_1 B^{m-1} + \cdots + b_m\,(= 0), \qquad b_i \in O_k$$

$$(6.1.6b) \quad g(A) = A^N + a_1 A^{N-1} + \cdots + a_N\,(= B), \qquad a_i \in O_k$$

then the chain rule $[df/dA = (df/dB)(dB/dA)]$ yields

$$(6.1.6c) \quad \delta_{K/k}(A) = \delta_{L/k}(B)\,\delta_{K/L}(A).$$

The representation of the fields by equations (6.1.6a, b) is actually special. It does work for function fields where the extensions are treated in terms of "local coordinates"; for example, $Z^e = z - \alpha$ or $1/z$. The use of so-called local methods in number theory is based in part on the desire to achieve this type of simplistic approach.

We must define the relative different and discriminant for an integer ring O_K over O_k for the extension K/k. To begin with a monogenic extension generated by $A \in K$,

$$(6.1.7a) \quad O_K = O_k[A],$$

we can take the ring different and ring discriminant to be the root different and root discriminant of A. There is the usual complication that not all rings are monogenic, but there also is a further problem that not all integer rings have a minimal relative basis:

$$(6.1.7b) \quad O_K = O_k[A_1, \ldots, A_n], \qquad [K:k] = n.$$

There is no difficulty in obtaining a nonminimal basis; for instance the absolute basis (over $\mathbf{Z}$) is also a (superfluously large) basis over O_k. A more reasonable basis theorem gives a result like equation (6.1.7b):

$$(6.1.7c) \quad O_K = O_k[A_1, \ldots, A_r]$$

where $r = n$ or $n + 1$ (but we do not need it for now). From here, we can define the discriminant as the square of the determinant:

$$(6.1.8) \quad d_{K/k} = D^2_{K/k}$$

where the determinant is now an *ideal* generated by the $R = r!/n!$ $(r - n)!$ $n \times r$ minors $A_{(i)}$ formed from the matrix of n (rows of) conjugates by r (columns of) basis elements A_j:

$$(6.1.9) \quad D_{K/k} = (\det A_{(1)}, \ldots, \det A_{(R)}).$$

This method, from Hilbert, is not as useful in practice as the following standardized definition (also from Hilbert):

6.1.10. Definition. *The relative "different" of K/k is defined by first finding certain ideals in $\hat{K}$ (the normal closure) called the "elements":*

$$(6.1.10a) \qquad E^{(t)} = (\ldots, A^{(1)} - A^{(t)}, \ldots), \qquad 2 \leq t \leq n$$

for the whole set of $A \in O_K$. The product is an ideal in K:

$$(6.1.10b) \qquad \delta_{K/k} = \prod_{t=2}^{n} E^{(t)}.$$

The relative "discriminant" is defined [see equations (6.1.3a, b)] as the ideal in k:

$$(6.1.10c) \qquad d_{K/k} = N_{K/k}\, \delta_{K/k}.$$

6.1.11. Theorem. *For the tower of fields $K \supseteq L \supseteq k$,*

$$(6.1.11a) \qquad \delta_{K/k} = \delta_{L/k}\, \delta_{K/L}$$
$$(6.1.11b) \qquad d_{K/k} = (N_{L/k}d_{K/L})d_{L/k}^{[K:L]}.$$

6.1.12. Remark. There is a third type of definition of different. With the usual notation, we define $\mathfrak{D}$ as a subset of K:

$$(6.1.12a) \qquad \mathfrak{D} = \{B \in K : Tr_{K/k}BA \in O_k, \text{ all } A \in O_K\}.$$

Because there is a finite (if not minimal) basis for O_K/O_k, there is only a finite number of A for which any B needs to be tested. Thus, the set $\mathfrak{D}$ is a fractional ideal in K (the difficult step is to check that the denominators are bounded). Also $\mathfrak{D} \supseteq O_K$ so $1/\mathfrak{D}$ is an integral ideal ($\mathfrak{D}$ divides O_K). We define

$$(6.1.12b) \qquad \delta_{K/k} = 1/\mathfrak{D}.$$

To verify the consistency of the definitions, it actually suffices to show this for the monogenic case $O_K = O_k[A]$, where A satisfies the monic equation of degree n, $f(A) = 0$. There,

$$(6.1.12c) \qquad \mathfrak{D} = [1, A, \ldots, A^{n-1}]O_k/f'(A).$$

Here, observing the multiplicity of definitions of discriminant, we see the advantages of monogenic integer ring extensions, where everything seems to agree so harmoniously (see Exercise 6.1). We shall use this idea locally, for each prime $\mathfrak{P}$ of K, as a basic method in Section 6.2.

6.2. Hilbert sequence of fields and groups

We consider the tower of fields L and integer rings O_L between K and k, where K/k is an extension of degree n:

(6.2.1a) $\qquad K \supseteq \cdots \supseteq L \supseteq \cdots \supseteq k$

(6.2.1b) $\qquad (O =) O_K \supseteq \cdots \supseteq O_L \supseteq \cdots \supseteq O_k (= o)$.

We view these extensions from the point of view of a prime ideal $\mathfrak{P}$ in the highest field. This determines a prime ideal $\mathfrak{P}_L$ divisible by $\mathfrak{P}$ in each of the fields in the tower. In particular, $\mathfrak{P} = \mathfrak{P}_K$ and we write $\mathfrak{p} = \mathfrak{P}_k$. We let p be the rational prime divisible by all these prime ideals. Thus, for all L (including K and k),

(6.2.1c) $\qquad \mathfrak{P}_L = \mathfrak{P}_K \cap O_L, \qquad \mathfrak{p} = \mathfrak{P}_L \cap \mathbf{Z}$.

We now ask precisely the same question of splitting that was asked in Chapter 2 in rational arithmetic of $K/\mathbf{Q}$: Does the prime ideal $\mathfrak{p}$ (of k) split into n distinct factors $\mathfrak{P}$ in K, and if not, how far can we go in terms of an intermediate L where splitting occurs? We redefine familiar symbols for order and degree:

(6.2.2a) $\qquad \text{ord } \mathfrak{P}/\mathfrak{p} = e \Leftrightarrow \mathfrak{P}^e \mid \mathfrak{p}, \qquad \mathfrak{P}^{e+1} \nmid \mathfrak{p}$

(6.2.2b) $\qquad \deg \mathfrak{P}/\mathfrak{p} = f \Leftrightarrow N_{K/k}\mathfrak{P} = \mathfrak{p}^f$.

We have (see Section 2.4) finite fields $o/\mathfrak{p}$ and its extension of degree f, $O/\mathfrak{P}$, where, in absolute terms,

(6.2.2c) $\qquad N[\mathfrak{p}] = |o/\mathfrak{p}| = p^s, \qquad N[\mathfrak{P}] = |O/\mathfrak{P}| = p^{fs}$

(6.2.2d) $\qquad \text{Gal } (O/\mathfrak{P})/(o/\mathfrak{p}) = C(f)$

a cyclic group of order f. This group is generated by

(6.2.2e) $\qquad X \to X^{p^s}, \qquad X \in O/\mathfrak{P}$

(note the power necessary to preserve $o/\mathfrak{p}$). Now "splitting" means $e = 1$ and $f = 1$. This leads to an equivalent statement of splitting.

6.2.3. Lemma. *A necessary and sufficient condition for $\mathfrak{p}$ of norm p^s to split from k to the prime $\mathfrak{P}$ in K is that $\mathfrak{p}$ be unramified at $\mathfrak{P}$ and*

(6.2.3a) $\qquad\qquad A^{p^s} \equiv A \bmod \mathfrak{P}$

for all A in O.

6.2.4. Lemma. *The order and degree are multiplicative:*

(6.2.4a) $\qquad\qquad$ ord $\mathfrak{P}/\mathfrak{p}$ = ord $\mathfrak{P}/\mathfrak{P}_L$ ord $\mathfrak{P}_L/\mathfrak{p}$

(6.2.4b) $\qquad\qquad$ deg $\mathfrak{P}/\mathfrak{p}$ = deg $\mathfrak{P}/\mathfrak{P}_L$ deg $\mathfrak{P}_L/\mathfrak{p}$.

We now restrict K/k to a normal extension. This makes K/L normal for each L in the tower but not L/k [unless some further condition holds, for instance, K/k (hence L/k) may be abelian].

From normality, if $\mathfrak{p}$ has factors $\mathfrak{P}_L^{(j)}$ in L for $j = 1, \ldots, g$,

$$\text{(6.2.5a)} \qquad \mathfrak{p} = \prod_{j=1}^{g} \mathfrak{P}_L^{(j)e}, \qquad N[\mathfrak{P}_L^{(j)}]^f = N[\mathfrak{p}]^f$$

(6.2.5b) $\qquad\qquad n = efg.$

We now consider $S(K)$ the whole lattice of fields $\{L : K \supseteq L \supseteq k\}$ as used in expression (6.2.1a). Clearly, $\mathfrak{p}$ splits from k to L when

(6.2.6a) $\qquad\qquad$ deg $\mathfrak{P}_L/\mathfrak{p}$ = 1

(6.2.6b) $\qquad\qquad$ ord $\mathfrak{P}_L/\mathfrak{p}$ = 1.

In particular, consider all such L; they form a sublattice $\mathcal{S}(Z)$ under inclusion (by an argument not unlike that of Corollary 2.1.13). Hence there is a maximal L in $\mathcal{S}(Z)$ called the "splitting" field of $\mathfrak{P}$ in K/k, designated by $K_Z^{\mathfrak{P}}$ (or merely K_Z if $\mathfrak{P}$ is understood).

Analogously, consider the sublattice $\mathcal{S}(T)$ of fields between K and k with the property that $L \in \mathcal{S}(T)$ when

(6.2.7a) $\qquad\qquad$ deg $\mathfrak{P}_L/\mathfrak{p}$ = $f_L \geq 1$

(6.2.7b) $\qquad\qquad$ ord $\mathfrak{P}_L/\mathfrak{p}$ = 1.

Thus, only the second condition is demanding; the prime $\mathfrak{p}$ need no longer be a splitting prime up to L; it may be of any degree ($\leq f$) but $\mathfrak{p}$ is unramified in L/k. Again in $\mathcal{S}(T)$ a maximal L exists (with $f_L = f$), the "inertial" field of $\mathfrak{P}$ in K/k, designated by $K_T^{\mathfrak{P}}$ (or merely K_T if $\mathfrak{P}$ is understood).

This maximality process can be performed again with $\mathcal{S}(V)$ defined as containing all L for which

(6.2.8a) $\qquad\qquad$ deg $\mathfrak{P}_L/\mathfrak{p}$ = $f_L \geq 1$

(6.2.8b) $\qquad\qquad$ ord $\mathfrak{P}_L/\mathfrak{p}$ = e_L with $(e_L, p) = 1$.

These fields are an inclusion lattice with a maximal field L called the "(first) ramification" field $K_V^{\mathfrak{P}}$ (or K_V if $\mathfrak{P}$ is understood). For this field $f_L = f$ and e_L is the part of e prime to p. (This part is called "tame" ramification; if the order e is divisible by p, the ramification is called "wild.")

An easy consequence is $\mathscr{S}(K) \supseteq \mathscr{S}(V) \supseteq \mathscr{S}(T) \supseteq \mathscr{S}(Z)$, or

$$(6.2.9) \qquad K \supseteq K_V \supseteq K_T \supseteq K_Z \supseteq k.$$

These lattice definitions are easy to state but not easy to work with. It is easier to define the sequence (named after Hilbert) by Galois group methods. Since K/k is normal, all intermediate fields L correspond biuniquely with G_L, the subgroups of $G = \text{Gal } K/k$. Thus,

$$(6.2.9a) \qquad G_L \leftarrow \text{Gal} \rightarrow L, \qquad G_L = \text{Gal } K/L$$

the Galois correspondence. We obtain G_L as the group fixing L:

$$(6.2.9b) \qquad G_L = \{U \in G : \alpha^U = \alpha \quad \text{for all } \alpha \in L\}.$$

We obtain L as the subfield invariant under G_L:

$$(6.2.9c) \qquad L = \{\alpha \in K : \alpha^U = \alpha \quad \text{for all } U \in G_L\}.$$

Hence, $G_K = 1$ (identity group) and $G_k = G$ (the whole group), and the relation is always contravariant:

$$(6.2.9d) \qquad L \supseteq L' \leftrightarrow G_L \subseteq G_{L'}.$$

The Hilbert sequence is defined by subgroups of G first. Here A is an arbitrary integer (in O_K). Consider

$$(6.2.10a) \quad G_Z = \{U \in G : \mathfrak{P}^U = \mathfrak{P} \quad \text{or} \quad A \equiv 0 \leftrightarrow A^U \equiv 0 \bmod \mathfrak{P}\}$$
$$(6.2.10b) \quad G_T = \{U \in G : A^U \equiv A \bmod \mathfrak{P}\} \quad (= G_{V_0})$$
$$(6.2.10c) \quad G_{V_r} = \{U \in G : A^U \equiv A \bmod \mathfrak{P}^{r+1}\}$$

(so $G_{V_1} = G_V$, and fields K_Z, K_T, K_{V_r}, and so on are labeled to correspond). It is next seen that because G_Z fixes $\mathfrak{P}$, then G_T, G_V, and so on are invariant subgroups of G_Z. Since G_Z preserves $\mathfrak{P}$, one of g conjugates,

$$(6.2.11a) \qquad |G/G_Z| = g.$$

Also, since G_T preserves each residue class mod $\mathfrak{P}$,

$$(6.2.11b) \qquad |G_Z/G_T| = |(O_K/\mathfrak{P})/(O_k/\mathfrak{p})| = |C(f)| = f$$

referring, of course, to the cyclic Galois group of an extension of a finite field. Furthermore,

$$(6.2.11c) \qquad |G_T| = e.$$

We have no direct need for further details on ramifications, but if $e = e_0 p^w$, where $(e_0, p) = 1$, then there is a cyclic quotient,

$$(6.2.11d) \qquad |G_T/G_{V_0}| = e_0$$

followed by further quotient groups of type $C(p) \times C(p) \times \cdots \times C(p)$,

$$(6.2.11e) \qquad G_{V_r}/G_{V_{r+1}} = p^{w_r} \qquad (w_r \geq 0, \ \Sigma w_r = w).$$

Of course, there is only a finite number of $w_r > 0$, indeed $p^w \mid n$.

In general, we deal with nonramified primes for which only G_Z and K_Z pose a problem. All higher fields are K, in particular,

$$(6.2.12) \qquad K_T = K, \qquad G_T = 1.$$

The ramified primes are necessary for many technical operations. We indicate their relation to the different and discriminant:

$$(6.2.13a) \qquad \delta_{K/k} = \prod_{\mathfrak{P}} \delta_{K/k}^{(\mathfrak{P})}, \qquad \delta_{K/k}(\mathfrak{P}) = \mathfrak{P}^{E_{\mathfrak{P}}}$$

$$(6.2.13b) \qquad d_{K/k} = \prod_{\mathfrak{p}} d_{K/k}^{(\mathfrak{p})}, \qquad d_{K/k}^{(\mathfrak{p})} = \mathfrak{p}^{E_{\mathfrak{P}} n/e}$$

where the exponent (for all $\mathfrak{P}$ in K that divide $\mathfrak{p}$ in k) is

$$(6.2.13c) \qquad E_{\mathfrak{P}} = e - 1 + \sum_r (p^{w_r} - 1).$$

If there is no wild ramification, $w = 0$, $K_V = K$, and then $E_{\mathfrak{P}} = e - 1$, reminiscent of the derivative of Z^e in local coordinates.

In the monogenic situation, $O_K = O_k[A]$, it suffices to compute the Hilbert sequence using only that generator A. This process can be generalized.

6.2.14. Definition. *The "local ring" mod $\mathfrak{P}$ is given by*

$$(6.2.14a) \qquad O_K^{\mathfrak{P}} = \{A/B : A, B \in O_K, \ B \not\equiv O \bmod \mathfrak{P}\}.$$

Such elements A/B are also called "$\mathfrak{P}$-integers."

6.2.15. Theorem. *The local ring $O_K^{\mathfrak{P}}$ is "monogenic by power series" over O_T, the integer ring for K_T for a given $\mathfrak{P}$. This means that a generator A_0 exists in the sense that for any $X \in O_K^{\mathfrak{P}}$, we can find an infinite sequence of $\gamma_j \in O_T$ so that for any $t > 0$,*

$$(6.2.15a) \qquad X \equiv \gamma_0 + \gamma_1 A_0 + \cdots + \gamma_t A_0^t \bmod \mathfrak{P}^{t+1}.$$

As much as we might like to do better, we cannot replace the base O_T by a lower-degree integer ring, since K_T is the first field where $\mathfrak{P}$ and $\mathfrak{P}_T$ have the same degree. The local ring is valuable for finding all the ramification fields and groups. The proof of the theorem requires the choice of any $A_0 = A/B$ so that $\mathfrak{P} \nmid B$ and $\mathfrak{P} \mid A$, $\mathfrak{P}^2 \nmid A$ (e.g., $B = 1$ for simplicity). Then we proceed step by step, choosing first $\gamma_0 \equiv X$ mod $\mathfrak{P}$. Next $\gamma_1 \equiv (X - \gamma_0)/A_0$ mod $\mathfrak{P}$, and so on. (This is the same way a Taylor series is created in step-by-step differentiation.)

The designation "local ring" means, generically, that the ring has only one maximal (prime) ideal, here (A_0). Actually all ideals are powers of the maximal ideal ("discreteness"). Certainly, these are properties that led inexorably to abstractionism in the form of "p-adic" numbers [infinite expansions like equation (6.2.15a)].

6.2.16. Remark. Kummer extension fields are defined as $K = k(\sqrt[l]{\mu})$, l prime ≥ 2, $\mu \notin k$, $\mu^{1/l} \notin k$. We assume k contains $L = \mathbf{Q}(\zeta)$ for $\zeta = \exp 2\pi i/l$ of degree $\phi(l) = l - 1$. Therefore,

$$G = \operatorname{Gal} K/k = \langle S : S^l = 1 \rangle, \qquad \sqrt[l]{\mu}^{\,S^t} = \zeta^t \sqrt[l]{\mu}$$

and for $(t, l) = 1$, $N_{L/\mathbf{Q}}(1 - \zeta^t) = l$, whereas for ϵ a suitable unit of L, $(1 - \zeta^t)^{l-1} = l\epsilon$ (see Exercise 6.2). We are interested in the ramification of a prime $\mathfrak{p}$ in O_k divided by $\mathfrak{P}$ in O_K (as usual, $O_k = o$, $O_K = O$). Since k is not necessarily factorial, we define for each $\mathfrak{p}$ the ideals $\mathfrak{q}$ and $\mathfrak{r}$ so that

$$\mathfrak{p}\mathfrak{q} = (\omega), \qquad \mathfrak{q}\mathfrak{r} = (\lambda), \qquad (N\mathfrak{p}, \lambda) = 1.$$

6.2.16a. Lemma. *By replacing μ by $\mu^x \xi^l$ (for ξ in K) we leave K unchanged; in particular, we can thereby replace μ by μ^* in o so that $(\mu^*, p) = o$ or $p \mid \mu^*$ and $p^2 \nmid \mu^*$.*

Proof. Begin by replacing $\mu = \alpha/\beta$ ($\alpha, \beta \in o$) by $\mu\beta^l = \alpha\beta^{l-1} \in o$. Write this new value as $(\mu) = \mathfrak{p}^e \mathfrak{a}$ ($\mathfrak{a}, \mathfrak{p}) = 1$. Then we can solve in $\mathbf{Z}$

$$(6.2.16b) \qquad\qquad ex - ly = g$$

where $x > 0$, $y > 0$ $(x, l) = 1$ and $g = 0$ if $l \mid e$, $g = 1$ if $l \nmid e$. We factor the ideal

$$(6.2.16c) \qquad (\mu)^x(\lambda)^{ly}/(\omega)^{ly} = \mathfrak{a}^x \mathfrak{p}^{ex}(\mathfrak{r}/\mathfrak{p})^{ly} = \mathfrak{a}^x \mathfrak{r}^{ly} \mathfrak{p}^g = (\mu^*)$$

where (μ^*) is now an integral ideal. If we replace μ^* by a suitable associate $\mu^*\epsilon$ ($\epsilon \in o^*$), we have $\mu^* = \mu^x(\lambda/\omega)^{yl}$.

Assume henceforth that the result of Lemma 6.2.16a is satisfied by a suitable replacement of μ.

Therefore [see equation (6.1.10a)] the element for K/k, $E^{(t)}$, contains the ideal $\sqrt[l]{\mu}(1 - \zeta^t)$ and the only ramified primes can be $\mathfrak{p} \mid \mu$ or $\mathfrak{p} \mid l$. For any ramified primes (by normality),

$$(6.2.16\text{d}) \qquad\qquad \mathfrak{p} = \mathfrak{P}^l \qquad (\mathfrak{P} = \mathfrak{P}^S).$$

Assume $\mathfrak{p} \mid \mu$ (independent of whether or not $\mathfrak{p} \mid l$), so

$$(6.2.16\text{e}) \qquad\qquad \mathfrak{P} = (\mathfrak{p}, \sqrt[l]{\mu}).$$

Since $O_T = o$ (deg $\mathfrak{P} =$ deg $\mathfrak{p}$), we have the "local" ring of $\mathfrak{P}$-integers:

$$(6.2.16\text{f}) \qquad\qquad O^{\mathfrak{P}} = o[\sqrt[l]{\mu}].$$

So $A_0 = \sqrt[l]{\mu}$ in Theorem 6.2.15 and A_0 replaces all A in equations (6.2.10a, b, c). Since $A_0 - A_0^S = (1 - \zeta) \sqrt[l]{\mu}$, we have only to determine

$$(6.2.16\text{g}) \qquad (1 - \zeta)\sqrt[l]{\mu} = \mathfrak{P}^E \mathfrak{g}, \qquad (\mathfrak{g}, \mathfrak{P}) = 1$$

so $E = 1$ if $\mathfrak{p} \mid \mu$ while $\mathfrak{p} \nmid l$, and $E > 1$ if $\mathfrak{p} \mid (\mu, l)$. Then by equations (6.2.10a, b, c),

$$(6.2.16\text{h}) \qquad \begin{aligned} (G_T =)G_{V_0} &= G_{V_1} = \cdots = G_{V_{E-1}} = G \\ G_{V_E} &= G_{V_{E+1}} = \cdots = 1 \end{aligned}$$

$$(6.2.16\text{i}) \qquad \delta_{K/k}^{(\mathfrak{P})} = \mathfrak{P}^{E(l-1)}, \qquad d_{K/k}^{(\mathfrak{p})} = \mathfrak{p}^{E(l-1)}. \qquad\qquad \square$$

When $\mathfrak{p} \nmid \mu$ but $\mathfrak{p} \mid l$ the ramification is rather complicated and breaks into many special cases; we consider just one.

6.2.17. Lemma (Kummer singular extension). *We specialize the preceding Kummer extension to the case where, for some integral ideal $\mathfrak{a}$ prime to l,*

$$(6.2.17\text{a}) \qquad\qquad (\mu) = \mathfrak{a}^l$$

and for some $\alpha \in o$,

$$(6.2.17\text{b}) \qquad\qquad \mu \equiv \alpha^l \bmod (1 - \zeta)^l$$

so $(1 - \zeta)^l = l(1 - \zeta)\epsilon_L$ for ϵ_L a suitable unit in $\mathbf{Q}(\zeta) = L$. Then K/k is unramified.

Proof. The only questionable prime is $\mathfrak{p} \mid l$, more precisely $\mathfrak{p} \mid (1 - \zeta)$, because $(1 - \zeta)^{l-1} = l$. We note, however, that

(6.2.17c) $\xi = (\alpha - \sqrt[l]{\mu})/(1 - \zeta)$ is monic.

Indeed, we observe that $N_{K/k}\xi = (\alpha^l - \mu)/(1 - \zeta)^l$ and

(6.2.17d)
$$\begin{aligned}(\xi(1 - \zeta) - \alpha)^l + \mu &= \xi(1 - \zeta)^l \\ &\quad + \{\text{terms in } l(1 - \zeta)\} - (\alpha^l - \mu) = 0.\end{aligned}$$

For ξ, the root different is clearly integral but there is no l-ramification.   □

An important related result is that where $l = 2$, $\zeta = -1$.

6.2.18. Lemma (Hilbert). *The field* $K = k(\sqrt{\mu})$, *for* μ *odd, has ramification only over odd primes when for some* $\alpha \in o$,

(6.2.18a) $\mu \equiv \alpha^2 \bmod 4$.

6.3. Classical illustrations

Here we offer a minimal number of classical illustrations of Hilbert sequences. The exercises will provide additional ones together with some supporting theory.

6.3.1. Quadratic field. *Let* $K = Q(\sqrt{d})$, $k = Q$, *and* $d = $ *discriminant. Here* $G = Gal\ K/k = \{1, U\}$, $\sqrt{d}^U = -\sqrt{d}$, *and* $O = Z[A]$ *for* $A = (d + \sqrt{d})/2$. *Therefore,* $A^U - A = -\sqrt{d}$. *Observe the following table. We list the primes* $\mathfrak{P}$ *in* K *on the right with relevant remarks:*

e	f	g	K_{V_3}	K_{V_2}	K_V	K_T	K_Z	k	$\mathfrak{P}$
1	1	2	K	K	K	K	K	Q	$\mathfrak{P}\mathfrak{P}^U = p$, $(\mathfrak{P} \neq \mathfrak{P}^U)$, $(d/p) = 1$
1	2	1	K	K	K	K	Q	Q	$\mathfrak{P} = p$, $(d/p) = -1$
2	1	1	K	K	K	Q	Q	Q	$\mathfrak{P}^2 = p\ (odd)$, $(\mathfrak{P} = \mathfrak{P}^U)$
2	1	1	K	K	Q	Q	Q	Q	$\mathfrak{P}^2 = p\ (even)$, $4 \mid d$, $8 \nmid d$
2	1	1	K	Q	Q	Q	Q	Q	$\mathfrak{P}^2 = p\ (even)$, $8 \mid d$.

Note that the primes decompose according to the Hilbert sequence.

6.3.2. Cyclotomic field of odd degree. *Here* $K = Q(\zeta)$ *for* $\zeta = exp\ 2\pi i/q$, *where* q *is an odd prime and* $k = Q$. *Then* $Gal\ K/k = \langle S \rangle$, *where* $\zeta^S = \zeta^r$ *for* r *a primitive root of* q. *The group* $\langle S \rangle$ *is cyclic of order* $q - 1$ $(= [K:Q])$. *If* $fg = q - 1$, *there is exactly one subgroup* G_f *of* G *of order* f *and exactly one subfield* K_g *of* K *of degree* g *where* $G_f \leftarrow Gal \rightarrow K_g$. *Actually,* $K_g = Q(\theta)$, *where*

(6.3.2a)
$$\theta = \sum_{t=1}^{f} \zeta^{r^t g}.$$

When $g = 2$, we have the Gaussian sum equal to $\sqrt{q^*}$ [$q^* = q(-1)^{(q-1)/2}$], so

(6.3.2b) $K_2 = \mathbf{Q}(\sqrt{q^*}).$

We show in Exercise 6.2 that $O = \mathbf{Z}[\zeta]$ and q is the only prime factor that divides the discriminant (q^{q-2}). This also implies that every intermediate field between $\mathbf{Q}$ and K must have a discriminant divisible only by q. Hence, alternatively, equation (6.3.2b) must be true.

We can show that if $p \nmid q$, then

(6.3.2c) $p = \mathfrak{P}_1 \cdots \mathfrak{P}_g, \quad N[\mathfrak{P}_j] = p^f, \quad p^f \equiv 1 \bmod q$

(and f is the minimum such positive integer). If $p = q$, then

(6.3.2d) $q = \mathfrak{Q}^{q-1} \text{ where } \mathfrak{Q} = (1 - \zeta), \quad N[\mathfrak{Q}] = q.$

The f and g are determined by factoring the generating polynomial for $\zeta \bmod p$ (where $p \nmid q$):

(6.3.2e) $(x^q - 1)/(x - 1) \equiv F_1(x) \cdots F_g(x) \bmod p$

into g polynomials $F_q(x)$ of degree f. Because $O/\mathfrak{P}$ has p^f elements, then by the finite field properties, $\zeta^{p^f} \equiv \zeta \bmod \mathfrak{P}$, which imples that $p^f \equiv 1 \bmod q$. Note that such an f must be minimal because ζ is a generator of the field and therefore if $p^{f'} \equiv 1 \bmod q$, then $O/\mathfrak{P}$ has no more than $p^{f'}$ elements.

We can summarize the Hilbert sequence as follows:

e	f	g	K_V	K_T	K_Z	k	$\mathfrak{P}$	
1	f	g	K	K	K_g	$\mathbf{Q}$	$\mathfrak{P} \mid p$, $(p, q) = 1$, $fg = q - 1$
$q - 1$	1	1	K	$\mathbf{Q}$	$\mathbf{Q}$	$\mathbf{Q}$	$\mathfrak{P} \mid q$

6.3.3. Quadratic reciprocity.

The maximal property of K_Z alone is deep enough to yield the following result.

6.3.3a. Lemma. *Using the decomposition law of equation (6.3.2c), g is even if and only if p splits in K_2.*

Proof. The "if" is easy because g is multiplicative (as if f). The hard part (the "only if") consists of showing that if p does not split in K_2, then it does not split in any higher subfield of K into an even number of factors. Since G is cyclic, all the subfields are partially ordered by divisibility of the degrees. Therefore, if p does not split in K_2, it does not split in any higher field of even degree. $\qquad\square$

6.3.3b. Quadratic reciprocity theorem.

$$(6.3.3c) \qquad (p/q) = 1 \Leftrightarrow p^{(q-1)/2} \equiv 1 \ mod \ q \Leftrightarrow \{g \ even\} \Leftrightarrow (q^*/p) = 1.$$

Proof. Note that $(q - 1)/2 = fg/2$, which is divisible by f precisely when g is even. (We use the minimal property of the degree f.) $\qquad\square$

6.3.4. Illustration for function fields. For $C(z, w) = K$ and $C(z) = k$ (K/k normal), the simple result is $f = 1$. This means that the rings $O/\mathfrak{P}$ and $o/\mathfrak{p}$ actually amount to the same set (the points C of the complex plane). The action of the Galois group is more complicated because it involves different paths around point α of the z-sphere along which the values of w [defined, say, by equation (5.1.1b)] are interchanged. Looking at the n sheets and n roots w_j locally, however, we see that they must break up into g ($= n/e$) cycles (or points) over $z = a$. The cycles are "rotated" and "interchanged" by the group. Actually,

$$(6.3.4a) \qquad K = K_V \supseteq K_T = K_Z \supseteq k.$$

The use of the local coordinate $z - \alpha = Z^e$ isolates the g points over $z = \alpha$ (as though $g = 1$), so the local group is merely cyclic of order e ($Z^S = Z \exp 2\pi i/e$, rotating the e values of Z cyclically). In other words, K_Z collapses to k. The fact that ramification is always tame follows from the fact that for the local coordinate, $Z - Z^S = Zc$ (c constant $\neq 0$) and is never divisible by Z^2, the condition (6.2.10c) for G_V.

The situation simplifies because C is algebraically closed. It would change drastically if the field of constants C were replaced by a finite field, for instance.

In conclusion, we make some remarks on terminology. First, despite any inclination to translate terms from German into English, the subscripts remain traditional:

$$Z = \text{Zerlegung (splitting)}$$
$$T = \text{Trägheit (inertia)}$$
$$V = \text{Verzweigung (ramification)}.$$

More significantly, the symbol is K_Z not k_Z, and so on because these fields are in principle closer to K. Actually, $G_Z = \text{Gal}(K/K_Z)$ is solvable, since its components have cyclic quotients [see equations (6.2.11b–e)]. Therefore, if G is unsolvable, $k \neq K_Z$ (and $g \neq 1$ so no prime p in k remains inert in K/k, for instance).

6.4. The Artin symbol

The residue classes $O/\mathfrak{P}$ for K form a field of $N[\mathfrak{P}] = p^{fs}$ elements. This is an extension of degree f over the residue classes $o/\mathfrak{p}$ for k that forms a field of $N[\mathfrak{p}] = p^s$ elements. The Frobenius automorphisms of $O/\mathfrak{P}$ over $o/\mathfrak{p}$ are cyclic of degree f generated by $A \to A^{p^s}$ (which preserves $o/\mathfrak{p}$). This observation leads to the following vital theorem of splitting fields.

6.4.1. Theorem. *There exists a unique element S in Gal K/k for any unramified prime ideal $\mathfrak{P}$ of K such that if $\mathfrak{P} \mid p$ (in k), then*

(6.4.1a) $$A^S \equiv A^{N[\mathfrak{p}]} \; mod \; \mathfrak{P}$$

for any A in O (indeed also for any $\mathfrak{P}$-integer of O). This S generates $G_Z^{\mathfrak{P}}$, cyclic of order $f \, (= deg_{K/k}\mathfrak{P})$.

The uniqueness proof consists of the observation that if $A^T \equiv A^U$ for T and U in G for all A in O, then TU^{-1} belongs to $G_T^{\mathfrak{P}}$, which is 1. (Recall that $\mathfrak{P}$ is unramified in K/k.)

6.4.1b. Definition. *The group element S in Theorem 6.4.1 is called the "Frobenius symbol" and written $[K/k/\mathfrak{P}]$.*

6.4.2. Conjugation lemma. *If $\mathfrak{P}$ is replaced by its conjugate $\mathfrak{P}^U$, then*

$$U^{-1}[K/k/\mathfrak{P}]U = [K/k/\mathfrak{P}U].$$

Therefore, if K/k is abelian, then the preceding symbol is independent of the particular $\mathfrak{P}$ that divides p and therefore is dependent on only p. Thus, equation (6.4.1a) is valid mod p as well as mod $\mathfrak{P}$.

6.4.3. Extension lemma. *Let $K \supseteq L \supseteq k$ be a tower with each extension assumed normal and let $\mathfrak{P}$ be an ideal in K unramified in K/k. Then*

(6.4.3a) $$[K/L/\mathfrak{P}] \equiv [K/k/\mathfrak{P}]^{[L:k]}.$$

Proof. Note that if $T = [K/k/\mathfrak{P}]$, then for all A in O_K, $A^T \equiv A^{N[\mathfrak{p}]}$ mod $\mathfrak{P}$ for $\mathfrak{p}$ in k and $\mathfrak{P}$ in K. Now, in the extension from k to L, $\mathfrak{p} \to \mathfrak{p}O_L$, $N[\mathfrak{p}] \to N[\mathfrak{p}]^{[L:k]}$, and $T \to T^{[L:k]}$. □

6.4.4. Definition. *For K/k abelian, the common value of* $[K/k/\mathfrak{P}^U]$ *for all U in G is called the "Artin symbol," written as* $(K/k/p)$. *Therefore, for A in O,*

(6.4.4a) $$A^{(K/k/p)} \equiv A^{N[p]} \ mod \ p.$$

This definition may be extended multiplicatively over all fractional ideas a in k whose prime factors (in k) are unramified. The process produces the Artin "mapping" of the ideal group generated by nonramified primes onto an image in G, written as

(6.4.4b) $$a \leftarrow A \rightarrow S.$$

In detail, if $a = \Pi p^e$, *then* $S = \Pi \ (K/k/p)^e$.

6.4.5. Quadratic illustration. Let $K = \mathbf{Q}(\sqrt{d})$, $k = \mathbf{Q}$, and Gal $K/k = \{1, U\}$, where $\sqrt{d}^U = -\sqrt{d}$. Let $p \nmid 2d$. Then, by Euler's lemma (2.2.11),

(6.4.5a) $$\sqrt{d}^p = \sqrt{d} \ d^{(p-1)/2} \equiv \sqrt{d}(d/p) \ mod \ p.$$

Note that all rationals r are preserved mod p under $r \rightarrow r^p$. Thus,

(6.4.5b) $\quad A = (r + s\sqrt{d})/2 \Rightarrow A^p \equiv [r + s\sqrt{d} \ (d/p)]/2 \ mod \ p.$

Clearly, $A^p = A$ or A^U, the conjugate (usually written A'). Therefore, $(K/k/p) = 1$ or U as $(d/p) = 1$ or -1. In this sense, we identify the symbols $(K/k/p)$ and (d/p) for $p \nmid 2d$. The kernel of the Artin mapping is the r in $\mathbf{Q}$ (relatively prime to $2d$ in numerator and denominator) for which $(d/r) = 1$. If we restrict r to integers in $\mathbf{Z}$, then r is a collection of arithmetic progressions.

6.4.6. Cyclotomic extensions. *Let* $k \supseteq \mathbf{Q}$ *and* $K = k(\zeta)$, *where* $\zeta = exp \ 2\pi i/m$. *Possibly* $k \cap \mathbf{Q}(\zeta) \supset \mathbf{Q}$ *so* $[K : k]$ *is a divisor of* $[\mathbf{Q}(\zeta) : \mathbf{Q}] = \phi(m)$ *and* $G = Gal \ K/k$. *Then, since* $K = k(\zeta)$,

(6.4.6a) $$(K/k/p) = U \in G$$

such that $\zeta^U \equiv \zeta^{p^s} \ mod \ p$ *where* p *does not ramify in* K/k *and* $N[p] = p^s$. *Thus* $(K/k/p)$ *is determined by* p^s *mod* m *in rational arithmetic. Easily, this symbol is determined by* p *mod* m. *For instance if* $N[p] \equiv 1$ *mod* m, *then* $(K/k/p) = 1$. *(Compare Exercise 4.7.)*

6.4.7. The image. *The Artin mapping covers all of* $G = Gal \ K/k$; *that is, G is generated by the* G_Z *of all unramified primes* $\mathfrak{P}$ *for K/k. To think otherwise would be to assume that* G_Z *generates a group* G_0 *properly* contained in G. Then $G_0 \leftarrow Gal \rightarrow K_0$, a subfield of K that properly contains k. This is absurd because all nonramified p would be splitting completely in K_0/k, contrary to the density theorem of Section 4.3. (The primes that split from **Q** to k could not have the same density as those that split from **Q** to K_0.)

6.4.8. The kernel. *We now come to the set of primes p in k for which the Artin symbol $(K/k/p) = 1$, the set that splits (completely) in K/k. The study of class fields will become the problem of characterizing these primes, as we shall see in Chapter 7. The previous examples give us a hint. In some sense, for K/k abelian, the splitting (nonramified) p in k lie in arithmetic progessions with a modulus built on (ramified) primes that are divisors of the relative discriminant.*

Exercises

6.1. Verify the equivalence of the three definitions of relative discriminant for the monogenic case $O = o[A]$. In particular, verify equation (6.1.12c) by driving the basis $\mathfrak{D} = [A_0, \ldots, A_{n-1}]o$ with $A_i \in K$, each chosen so its n conjugates satisfy the n equations ($j = 0, \ldots, n - 1$): $Tr\, A_i A_j = \delta_{ij}$.

6.2. Calculate the discriminant of $K/\mathbf{Q}$ for $K = \mathbf{Q}(\zeta)$, with $\zeta = \exp 2\pi i/p^t$ (p prime). Note that $\mathfrak{P} = (1 - \zeta)$ of norm p is the only ramified ideal and that all conjugates are the same. Also $O = \mathbf{Z}[1 - \zeta] = \mathbf{Z}[\zeta]$. Show that for p odd $d_{K/\mathbf{Q}} = p^E$, where $E = p^{t-1}t(p - 1) - 1$.

6.3. Composite biquadratic field. Let $K = \mathbf{Q}(\sqrt{d_1}, \sqrt{d_2})$, so K contains three quadratic subfields and $k_j = \mathbf{Q}(\sqrt{d_j})$ (d_3 is $d_1 d_2$ divided by a square). Consider only nonramified primes $\mathfrak{P}$ in K that divide p in **Q**. What are the residue conditions that determine K_z? (Note that no prime p is inert from **Q** to K.)

6.4. Cubic field. Let $k = \mathbf{Q}(\sqrt[3]{ab^2})$ for $(a, b) = 1$ be a cubic field and let k' and k'' be its conjugates (k is not normal). Now $L = \mathbf{Q}(\sqrt{-3})$ has the third root of unity and $K = kL = k'L = k''L$, a normal field. Consider again only nonramified primes in K; $\mathfrak{P} \mid p$ where $p \nmid 3ab$. Then there are three types of p distinguished by the residue symbol $(m/p)_3 = 1$, which means $x^3 \equiv m \bmod p$ is solvable for rational x:
1. $p \equiv \bmod 3$ and $(ab^2/p)_3 = 1$
2. $p \equiv \bmod 3$ and $(ab^2/p)_3 \neq 1$
3. $p \not\equiv 1 \bmod 3$.

These primes factor in k, L, and K as follows:

k	L	K
1. $p = \mathfrak{P}_{31}\mathfrak{P}_{32}\mathfrak{P}_{33}$	$p = \mathfrak{P}_{21}\mathfrak{P}_{22}$	$p = \mathfrak{P}_{61} \cdots \mathfrak{P}_{66}$
2. $p = p$	$p = \mathfrak{P}_{21}\mathfrak{P}_{22}$	$p = \mathfrak{P}_{21}\mathfrak{P}_{22}$
3. $p = \mathfrak{P}_{31}\mathfrak{P}_{30}$	$p = p$	$p = \mathfrak{P}_{31}\mathfrak{P}_{32}\mathfrak{P}_{33}$

(Note that k is not normal, so in case 3 the factors of p are not of the same degree; that is, $N[\mathfrak{P}_{30}] = p^2$, whereas $N[\mathfrak{P}_{31}] = p$.) These factorizations follow from Dedekind's theorem in Section 3.3 as applied to the defining equations for

$$k: \quad x^3 \equiv ab^2 \bmod p$$
$$L: \quad x^2 \equiv -3 \bmod p.$$

[If $p \not\equiv 1 \bmod 3$, the congruence for k always has just one solution, assuming $(p, 3ab) = 1$.] Find the Hilbert sequence for all unramified $\mathfrak{P}$ in K. (This means identifying only K_Z.)

6.5. In illustration 6.4.5 identify the Artin symbol with the quadratic residue symbol $(d/2)$ when d is odd ($p = 2$).

Bibliographic note

For general background, see Hilbert (1897, 1899), Hecke (1923), Hasse (1933), Lang (1970), and Weyl (1940). Special results on relative bases as in equation (6.1.7c) can be found in Narkewicz (1974, p. 24), but compare Hecke (1912). Additional results on cubic examples are based on Hasse (1930d).

7

The WHAT theorem of class field theory

At this stage we can define class field theory in a peremptory fashion by asking for the kernel of the Artin mapping. When we do so, a theorem of unusual elegance and power will result (see Section 7.4). It is called the WHAT theorem to commemorate Weber, Hilbert, Artin, and Takagi, a significant subset of the founders.

Merely stating this result would scarcely do justice to history; it would ignore the fact that class field theory emerged as a natural synthesis of number theory at a time when number theory was a synthesis of mathematics (and indeed when mathematics was a synthesis of science). Our approach, in turn, will not be strictly historical because we cannot afford to forgo the widom of hindsight by involving the reader in every significant but devious stage of development. An abridged version must suffice.

The two problems of number theory of prepossessing importance in the nineteenth century were associated with the two names Fermat and Gauss.

<table>
<tr><td align="center">*Fermat*</td><td align="center">*Gauss*</td></tr>
<tr><td>Fermat's last theorem (which we ignore here) required in some sense that fields such as $k = \mathbf{Q}(\exp 2\pi i/m)$ be imbedded in larger fields K so that all ideals in k become principal in K. (Of course, K can have nonprincipal ideals of its own.)</td><td>The theory of quadratic forms required conditions that primes in a field such as $k = \mathbf{Q}$ should split into two factors in a larger field such as $K = \mathbf{Q}(\sqrt{d})$.</td></tr>
</table>

Each of these requirements leads to a class field concept.

7.1. The two class field concepts

In the field extensions

$$K \supseteq k \, (\supseteq \mathbf{Q}), \qquad n = [K : k]$$

there are the integer rings,

$$O_K \supseteq O_k \, (\supseteq \mathbf{Z})$$

and ideals $\mathfrak{A} \subseteq O_K$ and $\mathfrak{a} \subseteq O_k$. According to the Fermat–Gauss dichotomy, two processes occur in the imbedding of k in K.

Principalization	*Factorization*
A nonprincipal ideal $\mathfrak{a}$ in k might become a principal ideal $(A) = AO_K$ in K by the imbedding $\mathfrak{a} = \mathfrak{a}O_k \to \mathfrak{a}O_K = (A)$	A prime ideal $\mathfrak{p}$ in k might no longer be prime in K. It is said to split (totally) when $\mathfrak{p} = \mathfrak{P}_1 \cdots \mathfrak{P}_n$ (distinct factors).

Then K is said to have a "class field property" over k because

the ideal class of $\mathfrak{a}$ is related to the question of whether or not $\mathfrak{a}O_K$ is principal in K (Kronecker's property).	the ideal class of $\mathfrak{p}$ is related to the question of how a prime $\mathfrak{p}$ will split from k to K (Weber's property).

These two properties were (somewhat prematurely) thought to be united in the Hilbert class field (also called the "absolute" class field).

7.1.1. Definition. *The "Hilbert" class field of k is the maximal unramified relatively abelian field k_H. There are two interpretations of k_H: It is called the "weak" class field if ∞ is also unramified (i.e., k_H and k are both real or both complex), otherwise (k real and k_H possibly complex) it is called the "strict" class field.*

7.1.2. Theorem. *For every field k a unique Hilbert class field k_H (strict or weak) exists for which Gal k_H/k is isomorphic with the respective class group. Thus the degree $[k_H : k] = h_+$ (or h), the appropriate class number.*

7.1.3. Theorem. *In the Hilbert class field,*

every ideal $\mathfrak{a}$ of k becomes principal in k_H (although other ideals of k_H may be nonprincipal, of course).	*the prime ideals $\mathfrak{p}$ of k that are principal (and only these) split completely from k to k_H. (We exclude primes $\mathfrak{p}$ ramifying from k to k_H.)*

These theorems, conjectured by Hilbert and proved by Furtwängler and Artin, constitute Hilbert's synthesis. Yet they have a major weakness consisting of an imbalance still unresolved today.

The principalization property does not completely determine k_H. There may be smaller unramified relatively abelian fields in which all ideals of k principalize. Thus it can not be true that all fields L with $K \subseteq L \subseteq K_H$ correspond biuniquely to subgroups of the class group.

The factorization property not only completely determines k_H but also establishes a biunique correspondence between all L with $k \subseteq L \subseteq K_H$ (unramified relatively abelian extensions L/k) and subgroups H of the class group of k; that is, $H \subseteq \mathrm{Cl}\{k\}$.

This might seem surprising in view of an a priori sense that principalization is simpler to explain than factorization. As before, let L_1 and L_2 denote fields between k and k_H, and let H_1 and H_2 denote subgroups of the class group $\mathrm{Cl}\{k\}$ (say, the strong class group). Then this is what happens:

If an ideal $\mathfrak{a}$ of k becomes principal in L_1, then it remains principal in every $L_2 \supseteq L_1$.

If the prime ideal $\mathfrak{p}$ of k splits completely in L_2, then it must have split completely in every $L_1 \supseteq L_2$.

Let us set up a "Hilbert" correspondence:

Let us set up a "Weber–Takagi" correspondence:

$$(7.1.4) \quad H_1 - \mathrm{H} \rightarrow L_1$$

$$H_1 - \mathrm{WT} \rightarrow L_1$$

to mean that the ideals in the subgroup H_1 all principalize in L_1 but in no smaller subfield. Now if $H_2 - \mathrm{H} \rightarrow L_2$, we have a covariance

to mean that the ideals in the subgroup H_1 all split completely in L_1 but in no larger subfield. Now if $H_2 - \mathrm{WT} \rightarrow L_2$ we have a contravariance

$$(7.1.5) \quad \begin{array}{c} H_1 \subseteq H_2 \\ \text{when } L_1 \subseteq L_2 \end{array}$$

$$H_1 \subseteq H_2 \text{ when } L_1 \supseteq L_2$$

only to the extent that the H-correspondence is unique (not always true).

unconditionally (by the uniqueness of the WT-correspondence).

7.1.6. Illustration of "good" principalization (biuniqueness) and "bad" principalization (failure of biuniqueness).

$$k = \mathbf{Q}(\sqrt{-195}), \quad h = 4, \quad \mathrm{Cl}\{k\} = \{1, 3_1, 5_1, 13_1\} = C(2) \times C(2)$$
$$k_H = \mathbf{Q}(\sqrt{-3}, \sqrt{5}, \sqrt{13}), \quad K_H \supset k_{4i} \supset k, \quad i = 1, 2, 3$$

$$\{1, 3_1\} \leftarrow\mathrm{H}\rightarrow k_{41} = \mathbf{Q}(\sqrt{-3}, \sqrt{65}). \quad \text{Only } 3_1 \text{ principalizes:}$$
$$(3, \sqrt{-195}) = (\sqrt{-3}).$$
$$\{1, 5_1\} \leftarrow\mathrm{H}\rightarrow k_{42} = \mathbf{Q}(\sqrt{5}, \sqrt{-39}). \quad \text{Only } 5_1 \text{ princializes:}$$
$$(5, \sqrt{-195}) = (\sqrt{5}).$$
$$\{1, 13_1\} \leftarrow\mathrm{H}\rightarrow k_{43} = \mathbf{Q}(\sqrt{13}, \sqrt{-15}). \quad \text{Only } 13_1 \text{ princializes:}$$
$$(13, \sqrt{-195}) = (\sqrt{13}).$$

The preceding illustration shows good principalization (see Exercise 7.1). By contrast, consider the following:

$$k = \mathbf{Q}(\sqrt{-21}), \quad h = 4, \quad \mathrm{Cl}\{k\} = \{1, 2_1, 3_1, 2_1 3_1\} = C(2) \times C(2)$$
$$k_H = \mathbf{Q}(i, \sqrt{-3}, \sqrt{-7}), \quad k_H \supset k_{4i} \supset k, \quad i = 1, 2, 3$$

In each of the intermediate fields $k_{41} = \mathbf{Q}(\sqrt{-3}, \sqrt{-7})$, $k_{42} = \mathbf{Q}(\sqrt{-7}, \sqrt{3})$, and $k_{43} = \mathbf{Q}(i, \sqrt{21})$, *all* the ideal classes principalize. Note that $2_1 = (3 + \sqrt{7})$ in k_{41}, $2_1 = (1 + \sqrt{3})$ in k_{42}, and $2_1 = (1 + i)$ in k_{43}, while $3_1 = (3 + \sqrt{21})/2$ or $\sqrt{3}$ or $\sqrt{-3}$ in the appropriate field. Thus, *each* subgroup H where $1 \subset H \subseteq \mathrm{Cl}\{k\}$ satisfies the relation $H \leftarrow\mathrm{H}\rightarrow k_{4i}$ for *each* $i = 1, 2, 3$. (See Theorem 8.3.11.)

The preceding illustration shows bad principalization. If we were to take the correspondence $H \leftarrow\mathrm{WT}\rightarrow L$ based on the splitting of primes instead, we would find that both cases lead to biuniqueness (both would be good).

From now on, we concentrate primarily on the factorization requirement for class fields. In addition to the biuniqueness of $H \leftrightarrow L$, we obtain a further advantage: By modifying the equivalence class concept to a narrower group than principal ideals (namely, "ray ideal" groups), we shall establish a fully biunique correspondence between *all* relatively abelian extensions L/k and subgroups of a new "ray class" group to be defined. In other words, we are no longer restricted to the abelian extensions of k that are unramified.

7.2. Ray class structure

7.2.1. Definition. *The semigroup of "extended" ideals (or divisors) $\tilde{m}$ in k is the formal product of "finite" integral ideals in O_k (written as m) and (possibly) "infinite" components (denoted by ˜) that correspond to the n (= [k/Q]) imbeddings of k and are denoted by ∞_i, where $i = 1, \ldots, n$. If all components occur, we abbreviate $\infty = \infty_1 \cdots \infty_n$. For* *an imaginary imbedding ($k^{(i)}$ complex), we write $\infty_i = 1$ (formally), but in any case $\infty_i \infty_i = \infty_i$. The concepts of gcd and lcm also apply formally if we examine the finite and infinite factors of each extended ideal.*

7.2.2. Definition. *Starting with k^*, the multiplicative subgroup of k, we define a subgroup*

$$k_{\tilde{m}}^* (= k_m^*) = \{\alpha/\beta : (\alpha\beta, m) = 1\}$$

where $\alpha, \beta \in O_k$ and $\beta \neq 0$. (Thus, the infinite factor in $\tilde{m}$ is ignored. Also an element of k may have several representations as α/β, but we require only one to satisfy the definition.)

7.2.3. Definition. *A multiplicative congruence $\mathrm{mod}^\times \tilde{m}$ is defined within the group $k_{\tilde{m}}^*$ to coincide with the usual (additive) congruence in O_k when m is finite in the following fashion:*

$$\alpha/\beta \equiv \gamma/\delta \ \mathrm{mod}^\times \tilde{m} \Leftrightarrow \alpha\delta \equiv \beta\gamma \ \mathrm{mod} \ m, \qquad \mathrm{sgn}(\alpha/\beta)^{(i)} = \mathrm{sgn}(\gamma/\delta)^{(i)}$$

for all (real) imbeddings $k^{(i)}$ where ∞_i is present as a factor of $\tilde{m}$. [We follow the convention that if $k^{(i)}$ is complex, then all nonzero elements of k are positive. Thus, $-1 > 0$ in the imaginary conjugate fields of $\mathbf{Q}(^3\sqrt{2})$ but not the real one.]

7.2.4. Definition. *The "unit ray" subgroup of $k_{\tilde{m}}^*$ is*

$$k_{\tilde{m}}^l = \{\alpha/\beta : \alpha/\beta \in k_{\tilde{m}}^*, \quad \alpha/\beta \equiv 1 \ \mathrm{mod}^\times \tilde{m}).$$

Thus, $k^* = k_l^l$. (The term "ray" comes from the idea of a half-infinite line segment.)

7.2.5. Definition. *Starting with J^*, the group of (fractional) ideals, we define the subgroup*

$$J_{\tilde{m}}^* (= J_m^*) = \{\mathfrak{a}/\mathfrak{b} : (\mathfrak{a}\mathfrak{b}, m) = 1\}$$

where $\mathfrak{a}$ or $\mathfrak{b}$ is a nonzero (finite) ideal of O_k. We further define the "unit ray ideal class" $\mathrm{mod}^\times m$ as the subgroup

$$J_{\tilde{m}}^l = \{(\alpha/\beta) : \alpha/\beta \in k_{\tilde{m}}^l\}.$$

[Note that an ideal (α/β) can have many representations as $(\epsilon\alpha/\beta)$ for a unit ϵ; it suffices to have only one where $\alpha/\beta \in k_{\tilde{m}}^l$.]

7.2.6. Definition. *We consider "unit ray class" structure $mod^\times \tilde{m}$ to mean that two ideals a and b in $J_{\tilde{m}}^*$ are equivalent when a/b is not only principal but also generated by an element of $k_{\tilde{m}}^1$. Thus,*

$$a \sim b \; mod^\times \tilde{m} \Leftrightarrow a/b \in J_{\tilde{m}}^1.$$

With this we have a class group

$$J_{\tilde{m}}^*/J_{\tilde{m}}^1 = Cl\{k, \tilde{m}\}, \qquad h_{\tilde{m}} = |Cl\{k, \tilde{m}\}|$$

generalizing $|Cl\{k, 1\}| = h$ and $|Cl\{k, \infty\}| = h_+$ for the weak and strong class structures.

7.2.7. Theorem. *The unit ray class number is given by*

$$h_{\tilde{m}}/h = \Phi[\tilde{m}]/[O^* : O^* \cap k_{\tilde{m}}^1]$$

where $\Phi[\tilde{m}] = \Phi[m]2^r$ for $\Phi[m]$, the Euler phi-function of m in k (the number of residue classes prime to m), and r is the number of real factors ∞_i in $\tilde{m}$. Note that $O^* \cap k_{\tilde{m}}^1$ is the set of units lying in $k_{\tilde{m}}^1$ (i.e., congruent to $1 \; mod^\times m$).

7.2.8. Illustration (quadratic field). For $n = 2$ and $\tilde{m} = \infty = \infty_1\infty_2$, $h_{\tilde{m}} = h = h_+$ if k is complex or if k is real and $\tilde{m} = \infty_1$ or ∞_2. If k is real and $\tilde{m} = \infty = \infty_1\infty_2$, then $\Phi[\tilde{m}] = 4$, while $O^* \cap k_{\tilde{m}}^1 = \langle \epsilon_+ \rangle$ for ϵ_+, the generator of totally positive units. Here $O^* = \langle -1, \epsilon_0 \rangle$ for ϵ_0, the fundamental unit. Thus, as $\epsilon_+ = \epsilon_0$ or $\epsilon_+ = \epsilon_0^2$, we have $h_+ (= h_\infty) = 2h$ or h. (Note that $[O^* : O^* \cap k_{\tilde{m}}^1] = 2$ or 4 accordingly.)

7.2.9. Illustration (arithmetic progression). For $n = 1$ and $k = \mathbf{Q}$, let $m > 0$ (in $\mathbf{Z}$). Then $k_{m\infty}^1 \cap \mathbf{Z}$ consists of the positive arithmetic progression $\{mx + 1\}$, where $x \geq 0$. Indeed,

$$J_{m\infty}^1 = \{(a/b) : a, b \, (\neq 0) \in \mathbf{Z}, \quad (ab, m) = 1,$$
$$a/b > 0, \quad a \equiv b \bmod m\}.$$

Thus, $J_{4\infty}^1$ includes the ideals (1), (5), (9), ... but not the negatively generated ideal (-3). On the other hand, J_4^1 contains (3) because (3) is also (-3). In other words, J_m^1 includes ideals from the whole progression $(mx + 1)$, where all $x \in \mathbf{Z}$, that generate the same *ideals* as the positive pair of progressions $(mx \pm 1)$, where $x > 0$. To apply Theorem 7.2.7, note that $\Phi[m\infty] = 2\phi(m)$ and $O^* = \{\pm 1\}$, so $O^* \cap k_{m\infty}^1 = \{1\}$ (canceling a factor of 2) to give $h_{m\infty}/h = \phi(m)$. Furthermore, the elements in $Cl\{\mathbf{Q}, m\infty\}$ correspond to the positive progressions $\{mx + a\}$ as a varies over the reduced residue classes module m.

7.3. The conductor

7.3.1. Definition. *We call the subgroup J of J^* a "ray ideal subgroup" $mod^{\times} \tilde{m}$ if*

$$J^l_{\tilde{m}} \subseteq J \subseteq J^*_{\tilde{m}}.$$

Each such J determines equivalence classes of ideals and a "ray class group"

(7.3.2) $$Cl\{k, \tilde{m}, J\} = J^*_{\tilde{m}}/J = (J^*_{\tilde{m}}/J^l_{\tilde{m}}) : (J/J^l_{\tilde{m}}).$$

We call $h(J)$ the "class number relative to group J," where $h(J)$ is the order of $Cl\{k, \tilde{m}, J\}$. Thus, $Cl\{k, \tilde{m}, J^l_{\tilde{m}}\} = Cl\{k, \tilde{m}\} = h_{\tilde{m}}$.

We note, however, that $|Cl(k, \tilde{m}, J^*)| = |Cl\{k, 1\}| = |h|$ by virtue of the fact that each ideal class has ideals prime to any preassigned modulus. Thus, the assignment of J is not completely unique; for example, $J^*_{\tilde{m}}$ "behaves" like J^* in the present context.

7.3.3. Definition. *Let J_1 be a ray ideal group $mod^{\times} \tilde{m}_1$ and let J_2 be a ray ideal group $mod^{\times} \tilde{m}_2$. Then we say that $J_1 \approx J_2$ (or J_1 and J_2 are "equivalent") if for some ideal m_0, a common intersection J_0 satisfies*

$$J_0 = J_1 \cap J^*_{m_0} = J_2 \cap J^*_{m_0}.$$

7.3.3a. Remark. We use the symbol J for any equivalent group as a matter of convenience in notation. Thus, we interpret the inclusion $J_1 \supseteq J_2$ somewhat loosely to mean that some groups J_{10} and J_{20} equivalent, respectively, to J_1 and J_2 satisfy $J_{10} \supseteq J_{20}$. Also, the only m_0 that enter into consideration (in this matter of equivalence) are divisors of $m_1 m_2$.

7.3.4. Lemma. *In the notation of Definition 7.3.3,*

$$Cl\{k, \tilde{m}_1, J_1\} \approx Cl\{k, \tilde{m}_2, J_2\} \approx Cl\{k, \tilde{m}, J_0\}$$

where $\tilde{m}$ is (say) the lcm of $\tilde{m}_1$ and $\tilde{m}_2$.

7.3.5. Definition. *The "conductor" $\tilde{f}$ of a ray ideal group J (or more precisely of its equivalence class) is the gcd of all $\tilde{m}$ of groups in the* *equivalence class. Thus, $\tilde{f}$ is the gcd of all $\tilde{m}$ for which the following statement is true:*

(7.3.5a) $\qquad J^l_{\tilde{m}} \subseteq J_0$ *for some J_0 equivalent to J.*

7.3.6. Definition. *We say two ideals are "congruent" $mod^\times$ m when their ratio is in $J^l_{\tilde{m}}$; that is,*

$$a \equiv b \leftrightarrow a/b \in J^l_{\tilde{m}}.$$

Thus, by abuse of notation, we write $a \equiv \beta$ for a an ideal and β an element of k to mean $a \equiv (\beta)$.

7.3.7. Remark. The conductor $\tilde{f}$ of J can also be defined as the gcd of all (extended) ideals $\tilde{g}$ such that

(7.3.7a) $\qquad\qquad a \equiv 1 \; mod^\times \; \tilde{g} \Rightarrow a \in J.$

Note that $\tilde{f}$ is the maximal of those $\tilde{g}$ (meaning it has the fewest factors).

7.3.8. Quadratic illustration. If $k = Q(\sqrt{d})$ and $d < 0$, then J^l_1 and J^l_∞ are equivalent (indeed identical). If $d > 0$, then J^l_1 and J^l_∞ are equivalent (and identical) if and only if there exists a unit ϵ_0 of norm -1. (In that case every principal ideal has a totally positive generator.) Less trivial cases appear in Exercise 7.8.

7.3.9. Rational illustration. If $k = Q$, then for odd $m > 1$, $J^l_{m\infty}$ and $J^l_{2m\infty}$ are equivalent (using $m_0 = 2$, for example). Less trivial examples appear in Exercise 7.2.

7.3.10. Remark. The nature of equivalence is that two equivalent groups may (as sets of ideals) differ in an infinite number of members, but they can differ in only a finite set of prime ideals (e.g., the divisors of m_0).

7.4. The WHAT theorem summarized

The field k has local and global attributes. Within k (i.e., locally) there is the ray class structure with class groups and conductor as obvious invariants. Outside of k (i.e., globally) there is the relatively abelian extension K with Gal K/k and $d_{K/k}$ as the obvious invariants. These correspond by the WHAT theorem (which comes in three parts: Theorems 7.4.1, 7.4.5, and 7.4.17).

7.4.1. WHAT theorem (Weber–Takagi). *For a number field k, there is a biunique WT-correspondence between (equivalence classes of) J, the ray ideal group in k, and K, the relatively abelian extension of k, written*

$$(7.4.2) \qquad\qquad J \leftarrow WT \rightarrow K.$$

This so-called Weber–Takagi correspondence is contravariant; that is, if $J_1 \leftarrow WT \rightarrow K_1$ and $J_2 \leftarrow WT \rightarrow K_2$, then

$$(7.4.3) \qquad\qquad J_1 \supseteq J_2 \Leftrightarrow K_1 \subseteq K_2.$$

In particular, there are (unit) ray class fields $K_{\tilde{m}}^1$ and groups $J_{\tilde{m}}^1$

$$(7.4.3a) \qquad\qquad K_{\tilde{m}}^1 \leftarrow WT \rightarrow J_{\tilde{m}}^1$$

such that (see Definition 7.2.5)

$$(7.4.3b) \qquad\qquad J_{\tilde{m}}^* \subseteq J \subseteq J_{\tilde{m}}^1 \Leftrightarrow k \subseteq K \subseteq K_{\tilde{m}}^1.$$

The essence of the WT-correspondence is that

$$(7.4.4) \qquad\qquad p \in J \Leftrightarrow p \text{ splits completely in } K/k \text{ (wfe)}$$

for p a prime ideal in k. If $\tilde{m}$ is maximal among equivalent J, that is, $\tilde{m} = \tilde{f}$ the conductor, then the relation (7.4.4) holds without exception. Thus, if J is given, K may be characterized as the maximal relatively abelian extension of k for which the following holds:

$$(7.4.4a) \qquad\qquad p \in J \Rightarrow p \text{ splits completely in } K/k \text{ (wfe)}.$$

Conversely, if K is given, then J and $\tilde{f}$ may be characterized as follows: For a variable (extended) ideal $\tilde{m}$ taken as a parameter and $\mathfrak{A}$ an arbitrary ideal of O_K prime to m, define

$$(7.4.4b) \qquad\qquad J[\tilde{m}] = \langle N_{K/k}\mathfrak{A}, J_{\tilde{m}}^1 \rangle$$

(the ideal group generated by the norms and ray ideal group shown). Then $\tilde{f}$ is the maximal (gcd) of all $\tilde{m}$ for which

$$(7.4.4c) \qquad\qquad p \in J[\tilde{m}] \Rightarrow p \text{ splits completely in } K/k \text{ (wfe)}$$

and $J = J[\tilde{f}]$.

The extension K/k has as discriminant divisors precisely those primes that divide the conductor.

7.4.5. The WHAT theorem (Artin). *Under the correspondence J $\leftarrow WT \rightarrow K$, there is an isomorphism of the Galois group of K/k and the ray class group, the so-called Artin isomorphism, as follows:*

$$(7.4.6) \qquad\qquad G = Gal\ K/k \leftarrow A \rightarrow Cl\{k, \tilde{f}, J\} = J_{\tilde{f}}^*/J.$$

Thus, in particular, the degree of the extension is the class number of the ray group,

(7.4.7) $n = |G| = [K/k] = |J_{\mathfrak{f}}^{*}/J| = h(J).$

There is, moreover, an explicit isomorphism from the Artin mapping. For $\mathfrak{a}$, a fractional ideal in J, we define (as in Section 5.4) the mapping

(7.4.8) $\mathfrak{a} \to (K/k/\mathfrak{a}) \in G.$

The kernel of the mapping is J [thus making equation (7.4.6) an isomorphism]. The mapping is defined for all $\mathfrak{a}$ without ramified prime divisors. For an arbitrary unramified prime $\mathfrak{p}$, $(K/k/\mathfrak{p})$ is an element of G of order n/g where $\mathfrak{p}$ splits into g factors $(\mathfrak{p} = \mathfrak{P}_1 \cdots \mathfrak{P}_g).$

7.4.9. Corollary. *The conductor of J, written $\tilde{\mathfrak{f}} = \tilde{\mathfrak{f}}(K/k)$, is characterized as the maximal ideal $\tilde{\mathfrak{m}}$ such that*

(7.4.10) $p \in J_{\tilde{\mathfrak{m}}}^{l} \Rightarrow p$ *splits completely in K/k*

or alternatively, for all $A \in O_K$,

(7.4.11) $p \in J_{\tilde{\mathfrak{m}}}^{l} \Rightarrow A \equiv A^{N[\mathfrak{p}]}$ *mod p.*

A similar characterization can be made for the finite part of the conductor, written $\mathfrak{f} = \mathfrak{f}(K/k)$.

7.4.12. Definition. *Let X denote the character group of G (= Gal K/k), with 1_X denoting the unit character. For any $\chi \in X$, there exists a subgroup of G said to "belong to χ," namely,*

$$G(\chi) = \{S \in G : \chi(S) = 1\}.$$

There also exists a field "belonging to χ," $K(\chi) \leftarrow Gal \to G(\chi)$, such that

$$K(\chi) = \{\alpha \in K : \chi(S) = 1 \Rightarrow \alpha = S\alpha\}$$

[the field of elements in K fixed by $G(\chi)$]. Finally, there is a conductor "belonging to χ," namely, $\tilde{\mathfrak{f}}(K(\chi)/k)$ [the conductor of $J(\chi)$ where $J(\chi) \leftarrow WT \to K(\chi)$]. Note that $G(1_X) = G \leftarrow Gal \to K(1_X) = k$, so $J(1_X) = J^*$ and $\tilde{\mathfrak{f}}(1_X) = 1$.

7.4.13. Conductor-discriminant theorem (Hasse).

$$d_{K/k} = \prod_{\chi \in X} \mathfrak{f}(\chi).$$

7.4.14. Illustration. For a cyclic extension K/k of prime degree l, $G(\chi) = 1$, $K(\chi) = K$, $J(\chi) = J(\leftarrow WT \to K)$, and $\mathfrak{f}(\chi) = \mathfrak{f}(K/k)$ for all characters except the principal 1_X. Thus, $d_{K/k} = \mathfrak{f}(K/k)^{l-1}$. For quadratic extensions over $\mathbf{Q}$, $(d) = \mathfrak{f}$. For a composite quadratic field, $K = \mathbf{Q}(\sqrt{d_1}, \sqrt{d_2})$, $k = \mathbf{Q}$, there are three quadratic subfields [the third is $\mathbf{Q}(\sqrt{d_1 d_2})$]. They correspond to three nontrivial characters. Thus, $d_{K/k}$ is the product of the three discriminants of the subfields.

7.4.14a. Remark. The fields $K(\chi)/k$ [and groups $G/G(\chi)$] are always cyclic (see Exercise 7.3).

7.4.15. Remark (reciprocity). The Artin correspondence is the quintessential expression of reciprocity in the classical sense of power (e.g., quadratic) reciprocity. This matter is too lengthy to pursue satisfactorily here. We merely point out that the WT-correspondence [relation (7.4.4)] puts the prime $\mathfrak{p}$ into opposite ("reciprocal") relationships. The statement "$\mathfrak{p} \in J$" is about the residue class of (a generator of) $\mathfrak{p}$ modulo $\tilde{\mathfrak{f}}$; the statement "p splits in K/k" is a statement of a polynomial (defining K) being factorable modulo p. This is like "$(p/d) = (d/p)$."

7.4.16. Remark (Hilbert class field). The two Hilbert class fields correspond to J_1^1 (weak) and J_∞^1 (strong) of relative degree h and h_+, respectively. By the WT-correspondence they are uniquely defined by the property that they are maximal relatively abelian and unramified (the weak one at ∞). They are also uniquely defined by the property that all principal ideals (in the strong case, with totally positive generators) will split (completely) from k to k_H. Unfortunately, the following more interesting property, outside the scope of this text, does not uniquely determine k_H.

7.4.17. WHAT theorem (Hilbert–Artin–Furtwängler). *The ideals of k all become principal in the (weak or strong) Hilbert class field k_H.*

The class field illustration par excellence is the absolutely abelian field K (abelian over $\mathbf{Q}$). Within $\mathbf{Q}$, if $\tilde{\mathfrak{f}}$ divides $m\infty$, then for $m > 1$, $J_m^* = \{(1), \dots, (m - 1)\}J_{m\infty}^1 \; (\sim J_1^*)$ [with $\phi(m)$ relative prime residue classes shown]. Hence, abelian fields correspond to special subgroups of residue classes mod m. These are the $J \leftarrow\text{WT}\rightarrow K$. Since $J_m^* \supseteq J \supseteq J_{m\infty}^1$, then by the WT-correspondence, if $J_{m\infty}^1 \leftarrow\text{WT}\rightarrow K_{m\infty}^1$ and $J \leftarrow\text{WT}\rightarrow K$, then $K \subseteq K_{m\infty}^1$.

7.4.18. Theorem (Kronecker). *All absolutely abelian fields K are generated by roots of unity, that is, are subfields of $\mathbf{Q}(\exp 2\pi i/m) \; (= K_{m\infty}^1)$,* where the conductor divides $m\infty$. A real abelian field K is a subfield of $\mathbf{Q}(\cos 2\pi/m)$ where m divides the (finite) conductor of K. In either case, the (gcd) minimal choice of all such $m\infty$ (or m in the real case) is the conductor.

Proof. We need only use the property of the conductor $m\infty$: Let $(p) \in J^1_{m\infty}$; that is, it splits completely in $K/\mathbf{Q}$ when $p \equiv 1 \bmod m$; hence, $K \subseteq \mathbf{Q}(\exp 2\pi i/m) = K^1_{m\infty}$. For real K, $(p) \in J^1_m$ instead and p splits completely in $K/\mathbf{Q}$ when $p \equiv \pm 1 \bmod m$; hence, $K \subseteq \mathbf{Q}(\cos 2\pi/m)$. The WT-correspondence does all the work! $\qquad\qquad\square$

7.4.18a. Corollary. *There are no unramified relatively abelian extensions of* $\mathbf{Q}$ *(Minkowski) or, for that matter, of any field k whose strict class number is unity. There are no unramified real relatively abelian extensions of any real field k whose weak class number is unity. (See Exercise 7.7.)*

7.5. Analytic class field theory

Finally, we view the local–global relationship analytically. The group of ideal classes $H = \mathrm{Cl}\{k, \tilde{\mathfrak{f}}, J\}$ is determined by the ray ideal group J (locally) and so is its character group $X = \{\chi(\mathfrak{a})\}$, now defined in terms of the ideals $\mathfrak{a}$ (with $\mathfrak{a} \in J^*_{\tilde{\mathfrak{f}}}$). There is, of course, a class field $K \leftarrow\text{WT}\rightarrow J$, with $G = \mathrm{Gal}\, K/k \sim H$, but it is a global aspect not yet evident. If $\mathfrak{f} = 1$, K can be ignored, but not if $\mathfrak{f} \neq 1$. In general, $\chi(\mathfrak{a})$ is definable also for some of the $\mathfrak{p}$ that divide $\mathfrak{f}$. [For instance, if $\chi = 1_X$, then $\chi(\mathfrak{a}) = 1$ for all $\mathfrak{a}$.] Thus, if we construct $K(\chi)$ between K and k and let $\mathfrak{f}(\chi)$ be its conductor, then $\chi(\mathfrak{a})$ is definable for additional $\mathfrak{a} \in J^*_{\mathfrak{f}(\chi)}$. Let us consider each $\chi(\mathfrak{a})$ definable for all (nonzero) ideals $\mathfrak{a}$ by extending the definition as stated, and by setting $\chi(\mathfrak{a}) = 0$ for $\mathfrak{a}$ not in $J^*_{\mathfrak{a}(\chi)}$. Call these the "extended" characters.

In Section 4.4 for an arbitrary ray class group H, with character group $\{\chi(\mathfrak{a})\}$ extended over all ideals $\mathfrak{a}$, we (essentially) defined the Hecke L-function:

$$(7.5.1) \quad L(s, k, \chi) = \sum_{\mathfrak{a}} \chi(\mathfrak{a})/N[\mathfrak{a}]^s = \prod_{\mathfrak{p}} (1 - \chi(\mathfrak{p})/N[\mathfrak{p}]^s)^{-1}$$

with $\mathrm{Re}\, s > 1$, $\mathfrak{a}$ is an ideal in O_k, and $\mathfrak{p}$ is a prime ideal in O_k.

7.5.2. Theorem. *The Dedekind zeta-function and the Hecke L-functions are related through class field theory by*

$$\zeta(s, K) = \prod_{\chi \in X} L(s, k, \chi).$$

Proof. First consider the case where $\mathfrak{f} = 1$. We regroup

(7.5.2a)
$$\prod_{\mathfrak{p}} = \prod_{\chi \in X} (1 - \chi(\mathfrak{p})/N[\mathfrak{p}]^s)^{-1} \quad (\mathfrak{p} \text{ fixed})$$
$$= \prod_{\chi \in X_{\mathfrak{p}}} \prod_{X/X_{\mathfrak{p}}} (1 - \chi(\mathfrak{p})/N[\mathfrak{p}]^s)^{-1}.$$

Here $X_{\mathfrak{p}}$ is the subgroup of X for which $\chi(\mathfrak{p}) = 1$ ($\mathfrak{p}$ fixed). Then $X/X_{\mathfrak{p}}$ is cyclic [by isomorphism with $G/G(\chi)$; see Remark 7.4.14a]. Its order is $f = n/g$, since by the Artin isomorphism, $\mathfrak{p}$ corresponds to an element $(K/k/\mathfrak{p})$ of order f when $\mathfrak{p} = \mathfrak{P} \dots \mathfrak{P}_g$ in K/k. (See Theorem 6.4.1.) Also $N[\mathfrak{P}_i] = N[\mathfrak{p}]^f$, so we have

(7.5.2b)
$$\prod_{X_{\mathfrak{p}}} \prod_{X/X_{\mathfrak{p}}} (1 - \chi(\mathfrak{p})z) = (1 - z^f)^g$$

identically in z, leading finally to

(7.5.2c)
$$\prod_{\mathfrak{p}} = \prod_{i=1}^{g} (1 - 1/N[\mathfrak{P}_i]^s)^{-1}.$$

The proof can be extended to the case where ramified primes occur by observing that the factor $\Pi_{\mathfrak{p}}$ is the same for k as for $K_{\mathfrak{P}}^T$, the inertial field (up to which $\mathfrak{p}$ is still unramified). But the degree of $X/X_{\mathfrak{p}}$ is now $f = n/(eg)$ because $K_{\mathfrak{p}}^T$ is of degree n/e. This is exactly what is required for equation (7.5.2c) to still hold.

7.5.3 Remark. The local aspect of the Hecke *L*-function may seem compromised when $\mathfrak{f} \neq 1$, but this is not really so. The identity in Theorem 7.5.2 is certainly true for nonramified primes, and there are analytic methods to show that such an identity valid for all but a finite number of primes is valid universally (in this context).

7.5.4. Theorem. *For character $\chi \neq 1_X$, $L(1, k, \chi) \neq 0$. For proof, note that $\zeta(s, k) = L(s, k, 1_X)$ and $\zeta(s, K)/\zeta(s, k)$ approaches a nonzero limit as $s \to 1$ (by the class number relations). Such $L(s, k, \chi)$ are analytic as $s \to 1$ and hence nonvanishing. (Compare Section 4.3, where the non-vanishing of the L-series implies that if the class field exists, it must have the correct degree h.)*

7.5.5. Theorem. *The primes in a given ideal class C (i.e., a given coset J_1^*/J) have Dirichlet density $1/h$;*

$$lim_{s \to 1} \left(\sum_{\mathfrak{p} \in C} N[\mathfrak{p}]^{-s} \right) \Bigg/ \left(\sum_{\mathfrak{p}} N[\mathfrak{p}]^{-s} \right) = 1/h.$$

Proof. Let C represent any ideal class. We are dealing with the asymptotic behavior as $s \to 1$ of the following:

$$\sum_{\mathfrak{p}} N[\mathfrak{p}]^{-s} \approx \log \mathfrak{z}(s, k) \approx \sum_{\chi} \log L(s, k, \chi)\chi^{-1}(C)$$
$$\approx \sum_{\chi} \sum_{\mathfrak{p}} \chi(\mathfrak{p})\chi^{-1}(C)/N[\mathfrak{p}]^s = h \sum_{\mathfrak{p} \in C} N[\mathfrak{p}]^{-s}$$

because the character sum over $\chi(\mathfrak{p}C^{-1})$ is h if $\mathfrak{p} \in C$ and 0 otherwise. There are infinitely many prime ideals in any ray ideal class. $\square$

7.5.6. Theorem. *The criteria involved in class field theory for the WT-correspondence may be restricted to primes of degree 1 and may be considered satisfied if valid except for a finite number of primes. For example, any two inequivalent ray ideal groups J_1 and J_2 with or without the same conductor will have an infinitude of primes of degree 1 that belong to one and not the other.*

7.5.7. Remark (Chebotarev density theorem). There is a stronger result that if K/K is normal but not necessarily abelian, then each Frobenius symbol $[K/k/\mathfrak{P}]$ for $\mathfrak{P}$ a prime in K is associated with a positive density of primes p of degree 1 in k divisible by $\mathfrak{P}$. We do not pursue these matters further, but we do need the result that there are infinitely many primes $\mathfrak{p}$ in k of degree 1 that have factors corresponding to a given Frobenius symbol. (Otherwise the WT-correspondence could hold vacuously because none of the required primes exists!) This density theorem actually arose in successive stages in the work of Chevotarev and Frobenius and Artin.

7.5.8. Illustration: Let $K = k(\exp 2\pi i/M)$ (see Exercise 4.7), so

$$K/k \leftarrow WT \to J = \langle \mathfrak{a} : N\mathfrak{a} = 1 \mod^{\times} M \rangle.$$

Note that we originally proved that if $N[\mathfrak{p}] = p$, then (wfe)

$$\mathfrak{p} \epsilon J \leftrightarrow N[\mathfrak{p}] = 1 \mod M$$

(but now, among other things, the condition that $N[\mathfrak{p}] = p$ is not needed).

Exercises

7.1. Complete Illustration 7.16 (on good principalization) by showing, for example, that $3 \neq 3_1^2$ for 3_1 principal in k_{42}. Try $3\epsilon = [(a + b\sqrt{5} + c\sqrt{-39} + d\sqrt{-195}/4]^2$ for ϵ a unit of k_{42}. It can be contradicted by seeing $3 \mid (a, b)$, so we are left with $\epsilon = (\alpha\sqrt{3} + \beta\sqrt{-13})^2/4$ for $\alpha, \beta \in O(\sqrt{5})$, which can be disproved.

7.2. Verify the following subfields of $\mathbf{Q}(\exp 2\pi i/12)$, that is, all abelian fields over $\mathbf{Q}$ whose conductors divide 12:

$$J_{12}^* = \{(1), (5), (7), (11)\}J_{12\infty}^1 \sim J_1^* \leftarrow WT \rightarrow \mathbf{Q}$$
$$\{(1), (7)\}J_{12\infty}^1 \sim J_{3\infty}^1 \leftarrow WT \rightarrow \mathbf{Q}(\sqrt{-3}) = \mathbf{Q}(\exp 2\pi i/3)$$
$$\{(1), (11)\}J_{12\infty}^1 \sim J_{12}^1 \leftarrow WT \rightarrow \mathbf{Q}(\sqrt{3}) = \mathbf{Q}(\cos 2\pi/3)$$
$$\{(1), (5)\}J_{12\infty}^1 \sim J_{4\infty}^1 \leftarrow WT \rightarrow \mathbf{Q}(i) = \mathbf{Q}(\exp 2\pi i/4)$$
$$\{(1)\}J_{12\infty}^1 \sim J_{12\infty}^1 \leftarrow WT \rightarrow \mathbf{Q}(\sqrt{3}, i) = \mathbf{Q}(\exp 2\pi i/12).$$

Do likewise for conductors that divide 20∞ and are from subfields of $\mathbf{Q}(\exp 2\pi i/20)$.

7.3. Prove that in the notation of Definition 7.4.12, $G/G(\chi)$ [hence $K(\chi)/k$] is always cyclic. Note that if $G = C(h_1) \times \ldots$ and $|G| = n = \Pi h_i$, then G is (additively) the vectors $S = \langle a_i, \ldots \rangle$ for $a_i \mod h_i$. Likewise, each χ is a vector $\langle b_i, \ldots \rangle$ for $b_i \mod h_i$. Thus, $\chi(S) = \exp 2\pi i c/n$, where $c = \Sigma a_i b_i n/h_i$. Then if S is fixed, the relation $\chi(S) = 1$ determines an additive subgroup of values of $c \mod n$. This gives a cyclic quotient.

7.4. Use Theorem 7.4.13 to derive the formula for $\mathfrak{f}(K/k)$ when Gal $K/k = C(4)$; namely, $\mathfrak{f}(K/k) = (d_{K/k}/d_{L/k})^{1/2}$ where L is the intermediate quadratic field. Check this for $k = \mathbf{Q}$, $L = \mathbf{Q}(\sqrt{2})$, and $K = \mathbf{Q}(\sqrt{2 + \sqrt{2}})$. Here we can check directly that $\mathfrak{f} = 16$, since $K = \mathbf{Q}(\cos 2\pi/16)$, $d_{L/k} = 8$, and, finally, $d_{K/k}$ can be found by the monogenic basis for $O_K = \mathbf{Z}[\zeta + 1/\zeta]$, where $\zeta^8 + 1 = 0$.

7.5. State a necessary and sufficient condition that the genus field of Chapter 2 is a Hilbert class field over the original quadratic field.

7.6. Show that for $p \equiv -1 \mod 8$, we can always find a solution to $x^2 - 2y^2 = -p$ with $4 \mid y$ [by using units in $\mathbf{Q}(\sqrt{2})$]. From this show that $K = \mathbf{Q}(\sqrt{-p}, \sqrt{2}, \sqrt{x + y\sqrt{2}})$ is unramified over $k = \mathbf{Q}(\sqrt{-2p})$. Draw conclusions about the class structure of k. Can this be done for $\mathbf{Q}(\sqrt{2p})$? (Try $p = 7$ first.)

7.7. Show that if a real quadratic field $\mathbf{Q}(\sqrt{d})$ has only the weak class number 1, then it has a relatively abelian unramified extension (which is, of course, complex). Show how to find it by trying several small values for d.

7.8. Show for $k = \mathbf{Q}(i)$ that J_2^1 and J^* are equivalent to each other but not to J_3^1. (For α odd in O_k, α or $i\alpha$ is $x + yi$ with y even.) Likewise show for $\mathbf{Q}(\sqrt{-3})$ that both J_2^1 and J_3^1 are equivalent to J^*.

Bibliographic note

For a historical survey, see Hasse (1967) and for technical details see Hasse (1926, 1927a, 1930b, 1933), Artin (1932), or Holzer (1966). An important historical source is Takagi (1920). For illustrations of the problems of principalization, see Taussky (1971). The conductor–discriminant theorem is found in Hecke (1917) and Hasse (1930c). Also see Artin (1924, 1927), Chebotarev (1926), Heilbronn (1967), Lagarias and Odlyzko (1977), Odlyzko (1977), and Weber (1899) for analytic results. For the Kronecker theorem, see Weber (1886) and Hilbert (1897).

8

The genus class field and transfer theory

For this exposition, the evolution of the concept of genus is a key to the theory. The original Gaussian concept of "genus" was (literally) a "type" of quadratic form. Two binary forms were to be of the same genus when they represent the same numbers to within square factors. Thus, this concept easily translates itself into classification by residue characters (see Chapter 2) and a class structure with equivalence by square factors (see Section 8.1).

This concept produces a genus field K_g over $k = \mathbf{Q}(\sqrt{d})$ as in Chapter 2, but now the fact of major importance is that K_g is *abelian* not only relative to k but also relative to $\mathbf{Q}$ (i.e., absolutely).

The role of genus field is seen in the context of a special type of ray class field called the "ring class field." To understand this concept, we think of what is called the "transfer" of class structures. We refer to the tower:

$$K \supseteq k \supseteq k_0$$

with each of the extensions K/k and k/k_0 abelian and K/k_0 normal. Then $K/k \; \leftarrow WT \rightarrow \; J$ and $k/k_0 \; \leftarrow WT \rightarrow \; j$, leading to two class structures:

$$H(J) = J_{\mathfrak{f}}^{*}/J \quad \text{and} \quad H(j) = j_{\mathfrak{f}_0}^{*}/j$$

for $\mathfrak{f} = \mathfrak{f}(k/k)$ and $\mathfrak{f}_0 = \mathfrak{f}(k/k_0)$. Then if $c \in H(j)$ is a class in k_0, we can select an ideal $\mathfrak{a}$ in c relatively prime to $\mathfrak{f}$ and $\mathfrak{f}_0$. This ideal becomes naturally imbedded in one of several possible ideal classes C in $H(J)$. We say that c "transfers" to the set of classes C.

Thus, we consider a Hilbert class field tower, with k_H the Hilbert class field of k and k_{HH} the Hilbert class field of k_H,

$$k \subseteq k_H \subseteq k_{HH}.$$

It has the interesting property (under Hilbert's part of the WHAT theorem 7.4.17) that with

$$k_H/k \leftarrow WT \rightarrow j \quad \text{and} \quad k_{HH}/k_H \leftarrow WT \rightarrow j_H,$$

all classes of $H(j)$ transfer into the principal class (namely, j_H of k_H). We now introduce a name for this property. If

$$\mathbf{Q} \subseteq k \subseteq K$$

then we shall *define* the "ring class fields" K over k by the property that if $j \leftarrow WT \rightarrow k/\mathbf{Q}$, then *all* classes of $H(j)$ transfer into $J \leftarrow WT \rightarrow K/k$.

The significance of this concept is hard to overestimate. Concrete applications of number theory and abstract concepts of group theory are linked by "transfer theory." In addition, the relation of Hilbert's class field concepts to Weber–Takagi theory is reasserted.

8.1. Ring class structure and genus

Let $K = \mathbf{Q}(\sqrt{d})$ as before and let some $f \in \mathbf{Z}^+$ be designated. We define the subring (or order f) in O_k as

(8.1.1)
$$O_f = \{\alpha \in O_k : \alpha \equiv z \bmod f\} \quad \text{for } z \in \mathbf{Z}.$$
$$O_f = [1, (df + f\sqrt{d})/2] \quad (O_1 = O_k).$$

(Historically, the word "order" for ring arose exactly in this connection.) There are ideals in O_f but they do not satisfy unique factorization because O_f is not integrally closed (only if $f = 1$). We do have unique factorization if we restrict the discussion to ideals relatively prime to f. They are called "regular" ideals of O_f.

8.1.2. Theorem. *Let a, $b \in J_1^*$ and $\in O_1$ (regular and integral). Then the operation on these ideals*

$$a_f = a \cap O_f, \qquad b_f = b \cap O_f$$

produces ideals in O_f with isomorphic structure; that is,

$$(ab)_f = a_f b_f.$$

This operation transforms the factorization into prime ideals in O_1 (= O) from p to p_f in O_f. To recover a from a_f, we define $a = a_f O_1$ (the module product). Thus,

$$a = [a, (b + \sqrt{d})/2], \qquad a_f = [a, f(b + \sqrt{d})/2].$$

Conversely, if $a_f = [a, (B + f\sqrt{d})/2]$, we solve $B \equiv bf$ mod $2a$ for b (using the fact that we have regular ideals). Then

$$a = a_f O_1 = [a, (b + \sqrt{d})/2].$$

We are dealing with a special ray class group now called a "ring class group." We replace the unit ray group J_f^1 by the "unit ring ideal group"

(8.1.3) $$J(\tilde{f}) = \{(z_1), (z_2), \ldots, (z_\phi)\} J_f^1$$

where z_i are the ϕ [$= \phi(f)$] reduced residue classes mod f. If $\tilde{f} = f\infty$, of course they are inequivalent, but if $\tilde{f} = f$, then $z_i/(f - z_i) \in J_f^1$ so there are only half as many. In general, we define a "ring ideal group" J as one that satisfies

(8.1.3a) $$J(\tilde{f}) \subseteq J \subseteq J_f^*.$$

8.1.4. Theorem. *A principal ideal (α) of O_1 that is regular remains principal in the structure of $J(f)$ if and only if $\alpha = \epsilon \alpha_f$ where $\epsilon \in O_1^*$ and $\alpha_f \in O_f$. We derive the "(unit) ring class group" and "(unit) ring class number" (as in Theorem 7.2.7) by taking the quotient group*

(8.1.5a)
$$\begin{cases} J_f^*/J(\tilde{f}) = H(\tilde{f}) & [= H(J(\tilde{f}))] \\ |H(\tilde{f})| = \Phi(\tilde{f})h/(\phi(f)[\tilde{O}^* : \tilde{O}^* \cap O_f]). \end{cases}$$

Here h is the ordinary class number of k, and $\tilde{O}^*$ represents the units restricted to the positive ray of $\tilde{f} = f\infty$. In more usual terms,

(8.1.5b)
$$\begin{cases} |H(f\infty)| = h_+(df^2) = h_+(d)f\Pi[1 - (d/p)/p]/E_+ \\ |H(f)| = h(df^2) = h(d)f\Pi[1 - (d/p)/p]/E \end{cases}$$

with products over the primes $p \mid f$. Here, when $d < 0$, $h = h_+$ and $E_+ = 1$, except $E_+ = 2$ or 3 for $d = -4$ and -3 (respectively). If $d > 0$, then E (or E_+) is the earliest power of ϵ_1, the fundamental unit (or ϵ_+, the fundamental totally positive unit) that lies in O_f.

There is a corresponding extension field K/k for $k = \mathbf{Q}(\sqrt{d})$,

(8.1.6a) $$[K(\tilde{f}) =] K \leftarrow\text{WT}\rightarrow J(\tilde{f})$$

called the "(unit) ring class field" for conductor $\tilde{f}$. For a (general) ring ideal group J, as in expression (8.1.3a), the WT-correspondence produces a (general) ring class field. This process is consistent with the generic definition of ring class field (in the introduction to this chapter). Note that the ideals (z) of $\mathbf{Q}$ transfer into ideals $(z)O_1$ in k and $J(\tilde{f})$ is generated as

$$(8.1.6b) \quad J(f) = \langle \mathfrak{a} : \mathfrak{a} \equiv z \bmod^{\times} \tilde{f}, \ z \in \mathbf{Z}, \ (z, fd) = 1 \rangle.$$

8.1.6c. Theorem. *The ring class fields K/k have the characterizing property that (wfe) the prime $\mathfrak{p}$ in k for which $\mathfrak{p} \equiv z \bmod^{\times} f$ $(z \in \mathbf{Z})$ are a subset of the primes that split from k to K. Furthermore, any intermediate subfield L $(K \supseteq L \supseteq k)$ is also a ring class field.*

The ring class group is of special importance because $H(f\infty)$ represents the class group of primitive quadratic forms of discriminant $f^2 d$. We set up two correspondences between modules and forms.

8.1.7. Definition. *If $\mathfrak{M}$ is a $\mathbf{Z}$-module in O_f, then the basis written $\mathfrak{M} = [\alpha, \beta]$ is said to be (positively) "oriented" if $(\alpha\beta' - \beta\alpha')/\sqrt{d} < 0$ (thus $[1, \sqrt{d}]$ is said to be oriented). Otherwise $\mathfrak{M}$ is said to have an "unoriented" (or negatively oriented) basis.*

Let $F(x, y) = ax^2 + bxy + cy^2$ be a binary form with $d_f = b^2 - 4ac = f^2 d$, $(a, b, c) = 1$, and $a > 0$ if $d < 0$. Here d is the field discriminant. Then $F \to \mathfrak{M}\{F\}$, an oriented module in O_f, where

$$(8.1.8a) \quad F \to \mathfrak{M}\{F\} = \begin{array}{ll} [a, (b + f\sqrt{d})/2] & \text{if } a > 0 \\ [a, (b + f\sqrt{d})/2]\sqrt{d} & \text{if } a < 0. \end{array}$$

The nature of the correspondence is such that it has an obvious inverse as follows: The norm form from $\mathfrak{M}\{F\}$ returns us to F. Specifically,

$$(8.1.8b) \quad \mathfrak{M} \to F\{\mathfrak{M}\} = N(\alpha x + \beta y)/t$$

where t is the (positive) gcd of the coefficients:

$$t = \gcd(\alpha\alpha', \alpha\beta' + \beta\alpha', \beta\beta') > 0.$$

Thus, $F\{\mathfrak{M}\{F\}\} = F$ with not even a loss of sign. On the other hand, $\mathfrak{M}\{F\{\mathfrak{M}\}\}$ is not always the same module as $\mathfrak{M}$. It would have an equivalent (oriented) basis to within a factor of a totally positive or totally negative number.

8.1.9. Lemma. *The correspondence (8.1.8a, b) is biunique betwen equivalence classes of primitive quadratic forms under a unimodular transformation for the discriminant df^2 and equivalence classes of modules in O_f under totally positive or negative multipliers (strict equivalence).*

The proof is an elementary use of the fact that the unimodular transformations of $F = F(x, y)$ correspond to the strict equivalence classes in $\mathfrak{M}$. Strangely enough, relaxing one type of strictness does not relax the other; if we permit transformations of the determinant ± 1 in (x, y), it does not correspond to weak equivalence of modules but rather to what we shall call "genus" equivalence. Observe that with the determinant ± 1, the conjugate classes C and C' are identified. Thus, C^2 is identified with CC', making square classes behave like principal classes.

8.1.9a. Remark. We must always bear in mind that although the correspondence of forms and ideals uses O_f, the ray or ring structure is in $O (= O_1)$ or J_f^* exclusively. (See Exercise 8.5 for the composition process for multiplying forms or ideals.) The broad scope of this theory is seen in the following lemma.

8.1.10. Lemma. *Every module $\mathfrak{M}$ in O is proportional to a regular ideal in some O_f.*

Proof. Construct $F\{\mathfrak{M}\}$ and change its basis so it has the form $F_0 = ax^2 + bxy + cy^2$ where $(a, f) = 1$. Then $\mathfrak{M}\{F_0\}$ is an ideal in O_f, from equation (8.1.8a), for $\beta = (b + f\sqrt{d})/2$ (d = field discriminant),

$$(8.1.10a) \qquad \mathfrak{M} = [a, \beta], \qquad O_f = [1, \beta] \qquad (\beta\mathfrak{M} \subseteq \mathfrak{M}). \qquad \square$$

8.1.11. Definition. *The "genus" ideal group $mod^\times f$ is defined as*

$$(8.1.11a) \qquad\qquad J_g(\tilde{f}) = J(\tilde{f})(J_f^*)^2.$$

Thus, two ideals $\mathfrak{a}$, $\mathfrak{b}$ of J_f^* are "genus equivalent" exactly when

$$(8.1.11b) \qquad \mathfrak{a} = \mathfrak{b}m^2\alpha/\beta, \qquad m \in J_f^*$$
$$\alpha, \beta \in O, \qquad \alpha/\beta \equiv z \; mod^\times \tilde{f}, \qquad (z, f) = 1, \qquad \beta \neq 0.$$

Otherwise expressed, the genus class group $mod^\times \tilde{f}$ is

$$(8.1.11c) \qquad\qquad H(J_g(\tilde{f})) = H(J(\tilde{f}))/H^2(J(\tilde{f})).$$

8.1.12. Lemma. *If $H(J(f)) = C(2^{t_1}) \times \cdots \times C(2^{t_g}) \times C(p^{t_p}) \times \cdots$ with g cycles representing powers of two and the other representing powers of odd primes (possibly repeated), then*

$$H(J_g(\tilde{f})) = C(2)^g.$$

8.1.13. Theorem. *The genus rank $\mathrm{mod}^\times f\infty$ is g, which denotes the number of Dirichlet characters $\chi_i(m)$ [= (d_i/m)] (see Section 2.3). This is the number of residue conditions that a positive prime p with $(p, 2m) = 1$ must satisfy to be represented by a form of order f that is equivalent to a perfect square (i.e., whose corresponding ideal class is a perfect square). If p is already represented by such a form, so $(d/p) = 1$, then one of the g conditions is superfluous.*

8.1.14. Theorem. *Two (primitive) binary quadratic forms $F_1(x, y)$ and $F_2(x, y)$ with discriminant df^2 correspond to the same genus $\mathrm{mod}^\times f\infty$ exactly when all numbers m_1, m_2 with $(m_1 m_2, f) = 1$ satisfy the condition that $F_1(x, y)$ represents $m_1 t_1^2$ exactly when $F_2(x, y)$ represents $m_2 t_2^2$ for suitable integers t_1 and t_2.*

8.1.15. Corollary. *Theorem 8.1.14 holds in a stronger sense that only one pair of integers m_1, m_2 is enough to determine equivalence.*

8.1.16. Remark. The key to Theorem 8.1.14 and Corollary 8.1.15 is that conjugate ideals are in the same genus; that is, $\mathfrak{p}\mathfrak{p}' = (p)$ and hence $\mathfrak{p}' = \mathfrak{p}(p/\mathfrak{p}^2)$. Likewise, two ideals with the same norm N are in the same genus [assuming $(N, 2f) = 1$].

8.1.17. Remark. The genus rank $\mathrm{mod}^\times f$ (under weak ideal equivalence) is possibly smaller than g when $d\ (> 0)$ has prime divisors congruent to $-1 \bmod 4$. (See Exercise 8.4.)

8.2. Absolute Galois structures

For an abelian extension K/k we know that Galois structure $G\ (= \mathrm{Gal}\ K/k)$ by its isomorphic image $H\ (= J_f^*/J)$, the ray class group structure in k. We now ask the further question: What can be said about the *absolute* Galois group $G_0 = \mathrm{Gal}\ K/\mathbf{Q}$ from the combined knowledge of G (i.e., H) and J? If we think of G as a subgroup of G_0, we note the Galois correspondence simultaneously with the Weber–Takagi correspondence as follows:

$$(8.2.1) \quad \begin{cases} 1 \xleftarrow{\text{Gal}} K, & K/k \xleftarrow{\text{WT}} J \bmod^\times \tilde{\mathfrak{f}} \\ G \xleftarrow{\text{Gal}} k, & k/\mathbf{Q} \xleftarrow{\text{WT}} j \bmod^\times \tilde{\mathfrak{f}}_0 \\ G_0 \xleftarrow{\text{Gal}} \mathbf{Q} \\ G_0 = \mathrm{Gal}\ K/\mathbf{Q}, & G = \mathrm{Gal}\ K/k \\ g = \mathrm{Gal}\ k/\mathbf{Q} \quad (= G_0/G \text{ for } K/\mathbf{Q} \text{ normal}). \end{cases}$$

The respective degrees are

(8.2.1a) $|G| = [K:k] = n,$ $|g| = [k:\mathbf{Q}] = m,$
$|G_0| = [K:\mathbf{Q}] = mn.$

In general, $K/\mathbf{Q}$ is said to "split" at an intermediate field k ($K \supseteq k \supseteq \mathbf{Q}$) when another intermediate field K_0 exists ($K \supseteq K_0 \supseteq \mathbf{Q}$) such that

$$K = K_0 \times k, \qquad [K_0:\mathbf{Q}] = n, \qquad [K:K_0] = m, \qquad (K_0 \cap k = \mathbf{Q}).$$

To describe the groups somewhat differently, whenever G is a normal subgroup of G_0 we can represent the quotient G_0/G as

(8.2.1b) $g = \{GT_0, GT_1, \ldots, GT_{m-1}\}, \qquad (T_0 = 1)$

a collection of cosets with $T_i \in G_0$ and a well-defined correspondence $(i, j) \to k$ that governs $GT_iGT_j = GT_k$, the multiplication in g. This does not mean that $T_iT_j = T_k$ always (unless g is injective in G_0). Indeed, when g is abelian, it means only that $GT_iGT_j = GT_jGT_i$, and not even that $T_iT_j = T_jT_i$. When g is injective in G, splitting occurs and $K_0 \leftarrow \text{Gal} \rightarrow g$.

8.2.1c. Remark. If g is cyclic of order m, we can choose representatives so that $T_i = T^i$ for some $T \in G_0$ ($T \notin G$), where $T^m \in G$, but no earlier power. Of course, $T^m = 1$ only when g is injective, but at any rate the elements in $\{T_i\}$ are now a commutative subset of G_0.

Clearly, the splitting of K at k is a property of the absolute Galois group structure. The following is a very early classic theorem on how this is related to the class field concept.

8.2.2. Monodromy theorem of Chebotarev. *If $K/\mathbf{Q}$ is normal, then $G_0 = \text{Gal } K/\mathbf{Q}$ is generated by the union of inertial Hilbert subgroups $G_T^{\mathfrak{P}}$ for $\mathfrak{P}$ that are ramified in $K/\mathbf{Q}$.*

Proof. Let $G^* = \langle G_T^{\mathfrak{P}} : \mathfrak{P} \text{ ramified} \rangle$. Thus, $G^* \supseteq G_T^{\mathfrak{P}}$ for each $\mathfrak{P}$. Hence, if $G^* \leftarrow \text{Gal} \rightarrow K^*$, then $K^* \subseteq K_T^{\mathfrak{P}}$ and each p (in $\mathbf{Q}$, where $\mathfrak{P} \mid p$) remains unramified up to K^*. By Corollary 4.1.14a to Minkowski's theorem, $K^* = \mathbf{Q}$, so $G^* = G_0$.   □

8.2.2a. Corollary. *If $K \supseteq k \supseteq \mathbf{Q}$ with K/k unramified, $K/\mathbf{Q}$ normal, and $k/\mathbf{Q}$ (cyclic) of prime order m, then G_0 (= Gal $K/\mathbf{Q}$) can be generated by elements of order m only, and K splits at k.*

Proof. By the preceding argument all ramifications occur in $k/\mathbf{Q}$, and therefore $G_{\mathfrak{T}}^{\mathfrak{P}} = m$ (or 1) for all $\mathfrak{P}$. So G_0 is generated by elements of order m. There must be at least one generator $T \notin G$ because $G_0 \neq G$. Thus, $g = \langle T : T^m = 1 \rangle$ is imbedded in G_0. □

8.2.3. Artin isomorphism theorem. *Let $K \supseteq k \supseteq \mathbf{Q}$ with $K/\mathbf{Q}$ normal, K/k abelian, and $k/\mathbf{Q}$ normal. Then if $H = J_f^*/J$ (the group of classes $\{C\}$ for $K/k \leftarrow WT \rightarrow J$), H is invariant under $g = \mathrm{Gal}\ k/\mathbf{Q}$ in the sense that*

(8.2.4a) $\qquad C \leftarrow A \rightarrow S \Leftrightarrow C^t \leftarrow A \rightarrow S^t, \quad t \in g, \quad S \in G.$

Here t is defined by the coset structure $\{GT_i\} = G_0/G$ $(t = GT)$, so $C' = C^{GT} = C^T$ while $S^t = T^{-1}ST$ $(= S^T)$, where S^t is uniquely defined on each coset, since G is abelian. [See equation (8.2.1b).]

Proof. This is a direct consequence of the Artin symbol. We use the more general notation $\mathbf{Z}[g]$ for the "group ring" of g. Thus, if g is cyclic of prime order m, then $\mathbf{Z}[g] = \{a_0 + a_1 t + \cdots a_{m-1} t^{m-1}\}$, where $a_i \in \mathbf{Z}$. Accordingly, with $x \in \mathbf{Z}[g]$,

(8.2.4b) $\qquad\qquad C \leftarrow A \rightarrow S \Leftrightarrow C^x \leftarrow A \rightarrow S^x.$

In particular, C^{1-t} means C/C^t. (Theorem 8.2.3 does not depend on K splitting at k, of course, or on g being cyclic.) □

8.2.5. Theorem. *Let K/k be abelian and $K/k \leftarrow WT \rightarrow J$. Then $K/\mathbf{Q}$ is normal precisely when J is invariant under g. Also if g is cyclic and $K/\mathbf{Q}$ normal, then $K/\mathbf{Q}$ is abelian precisely when not just J but all ideal classes in $H = J_f^*/J$ are invariant under g.*

Proof. The first part follows directly from the isomorphism. If K is invariant under g, then $K/k \leftarrow WT \rightarrow J \Leftrightarrow K^t/k \leftarrow WT \rightarrow J^t$ $(t \in g)$. Thus, $K = K^t \Leftrightarrow J = J^t$. The second part follows from Theorem 8.2.6. □

8.2.6. Theorem. *Let $K/\mathbf{Q}$ be normal with K/k abelian and $k/\mathbf{Q}$ cyclic, so that $K/k \leftarrow WT \rightarrow J$. Then the maximal intermediate field K_g $(K \supseteq K_g \supseteq k)$ that is abelian over $\mathbf{Q}$ is determined by $J_g \leftarrow WT \rightarrow K_g/k$, where J_g is the subgroup of the class group H generated by C^{1-t} $(t \in g$ and $C \in H = J_f^*/J)$.*

Proof. Here K_g corresponds to the commutator subgroup of G_0. It is generated by commutators $K = U_1 U_2 U_1^{-1} U_2^{-1}$ where $U_i \in G_0$. If we write

$U_i = S_iT_i$ ($i = 1, 2$; $S_i \in G$; $GT_i \in g$), then the S_i commute because G is abelian and the T_i commute because g is cyclic. Thus, we have

$$\begin{aligned}
[U_1, U_2] &= (S_1T_1)(S_2T_2)(S_1T_1)^{-1}(S_2T_2)^{-1} \\
&= S_1(T_1S_2T_1^{-1})T_1T_2T_1^{-1}T_2^{-1}(T_2S_1^{-1}T_2^{-1})S_2^{-1} \\
&= S_1S_2^{T_1-1}S_1^{-T_2-1}S_2^{-1} \\
&= S_1^{1-T_2-1}/S_2^{1-T_1-1} \leftarrow A \rightarrow C_1^{1-t_2-1}/C_2^{1-t_1-1}. \qquad \square
\end{aligned}$$

8.2.7. Remark. For $k = \mathbf{Q}(\sqrt{d})$ and $g = \langle t : t^2 = 1 \rangle$, $C^t = C^{-1}$ so J_g is generated by C^2 for $C \in H$. Thus, the field K_g under theorem 8.2.6 generalizes the genus concept to the new definition: The genus field for the abelian extension K/k within $K/\mathbf{Q}$ is the maximal intermediate field K_g abelian over $\mathbf{Q}$. If only k is given, as is often the case, we assume K refers to the (strict or narrow) Hilbert class field and K_g is called the (appropriate) Hilbert genus field of k.

We next seek to characterize ring class fields by special properties. For the remainder of Section 8.2, K/k is abelian of degree n (>1), $K/\mathbf{Q}$ is normal of degree nm, and k is quadratic ($m = 2$).

8.2.8. Lemma. *If K/k is a ring class field, then $K/\mathbf{Q}$ is never cyclic.*

Proof. Note that $K/\mathbf{Q}$ is cyclic only if $S = (K/\mathbf{Q}/(p))$, the Artin symbol, can be of order $2n$ in G_0 for some prime p. We show that this happens for no (unramified) p. For any such p, the ideal (p) would have to be inert from $\mathbf{Q}$ to K and hence surely from $\mathbf{Q}$ to k. Then, since $(p) \in J(f\infty)$ in k, we have a contradiction as (p) then splits from k to K (by the ring class field property). $\qquad \square$

8.2.9. Lemma. *Let $G_0 = C(2)^{r+1}$, so $K = \mathbf{Q}(\sqrt{d_0}, \ldots, \sqrt{d_r}) \supseteq k = \mathbf{Q}(\sqrt{d})$. Then K is a ring class field over k.*

Proof. Let $f = d_0 \cdots d_r$ and let $p = \mathfrak{p}\mathfrak{p}'$ for $(p, f) = 1$ (in k). Then the statement $\mathfrak{p} \equiv a \bmod^\times f$ implies that $\mathfrak{p}\mathfrak{p}' = (p) \equiv a^2 \bmod^\times f\infty$, so p splits in all fields $\mathbf{Q}(\sqrt{d_i})$, hence K. If, however, $p = (p)$ in k (p does not split in $k/\mathbf{Q}$), then p must surely split in K/k at each of the $(r - 1)$ adjunctions of order 2. (Note that by Theorem 7.4.2 it is not really necessary to consider such p.) The assumption $\mathfrak{p} \in J(f\infty)$ implies (wfe) that $\mathfrak{p}$ splits from k to K. $\qquad \square$

These lemmas lead to the following results:

8.2.10. Lemma. *Let $G = C(2)$ and $g = C(2)$ also, so $k = \mathbf{Q}(\sqrt{d})$. Then G_0 is of order 4, leading to two cases:*
1. *$G_0 = C(2) \times C(2)$; $K = \mathbf{Q}(\sqrt{d}, \sqrt{d_1})$ (ring class field),*
2. *$G_0 = C(4)$; $K = \mathbf{Q}(\sqrt{(a + b\sqrt{d})})$ with $a^2 - b^2 = c^2 d$ (a, b, c $\in$ **Z**)   (not a ring class field).*

8.2.11. Lemma. *Let $g = C(2) = \langle t : t^2 = 1 \rangle$. Then K/k is a ring class field only if for every class C in $H(J)$, $CC^t = 1$.*

Proof. If $\mathfrak{a}$ is an ideal in C, then $\mathfrak{a}\mathfrak{a}'$ is an ideal from **Z**.   □

8.2.12. Lemma. *Under the preceding hypotheses, with K/k also of odd degree (and abelian), $CC^t = 1 \Rightarrow J \supseteq J(f\infty)$.*

Proof. Let $\mathfrak{f}(K/k) = \mathfrak{f}$ and let $\mathfrak{f} \mid f \in \mathbf{Z}^+$ (e.g., $f = N[\mathfrak{f}]$). Then let $(a, f) = 1$, where $a \in \mathbf{Z}$. Since $(a) = (a)^t$, if $(a) \in C$, then $C = C^t$ so $C^2 = 1$, which implies $C = 1$ (odd order).   □

Lemmas 8.2.11 and 8.2.12 lead to a special result for $n = 3$. There $G = \langle C : C^3 = 1 \rangle$ so $C^t = C^{-1}$.

8.2.13. Theorem. *Let $H = C(3) = \langle C : C^3 = 1 \rangle$. Then either*
1. *$C^t = C^{-1} \Leftrightarrow J \supseteq J(f\infty)$ for some $f \in \mathbf{Z}^+$ (dihedral of order 6 and ring class field) or*
2. *$C^t = C \Leftrightarrow J \supseteq J(f\infty)$ for any $f \in \mathbf{Z}^+$ (abelian of order 6, hence cyclic, and not a ring class field).*

8.2.14. Illustration. If $k = \mathbf{Q}(\sqrt{-3})$ and $K = k(\sqrt[3]{a^2 b})$, $(a, b) = 1$, the pure cubic normal field, then $f = 3ab$ if $a^2 \not\equiv b^2 \bmod 9$ and $f = ab$ if $a^2 \equiv b^2 \bmod 9$. Here J is a ring ideal group $\bmod^\times f$ corresponding to a subgroup of index 3 in the group of quadratic forms of discriminant $-3f^2$.

8.2.15. Gauss–Dirichlet genus theorem. *Consider the (strict) equivalence classes of proper ideals of O_f or of forms of order f for $k = \mathbf{Q}(\sqrt{d})$. Those classes that are perfect squares are precisely those with forms whose Dirichlet characters are 1.*

8.2.16. Lemma. *There exists a finite set of discriminants $d_1, \ldots, d_{\hat{g}}$ such that if $p \nmid 2fd$, then for $F(x, y)$, the principal form of discriminant df^2,*

$$(d_i/p) = 1, \qquad i = 1, \ldots, \hat{g} \Leftrightarrow pt^2 = F(x, y)$$

(i.e., the representation is valid in terms of integral x, y, t).

Proof. The square ideal classes correspond to the genus field $K_g \supseteq k \supseteq \mathbf{Q}$ and $G = \text{Gal } K_g/k = C(2)^{\hat{g}}$ for some (as yet unknown) $\hat{g}$. But $K_g/\mathbf{Q}$ is also abelian ($C^{1-\iota} = C^2$), so $G_0 = \text{Gal } K_g/\mathbf{Q}$ is either $C(2)^{\hat{g}} \times C(4)$ or $C(2)^{\hat{g}+1}$. The $C(4)$ simply cannot occur because then $G_1 = C(4)$ $\leftarrow$Gal$\rightarrow K'$ where $[K' : \mathbf{Q}] = 4$ and K' as a subfield of a ring class field is still a ring class field, in contradiction to Lemma 8.2.8. Thus, $K_g = \mathbf{Q}(\sqrt{d_1}, \ldots, \sqrt{d_{\hat{g}+1}})$ corresponding to $C(2)^{\hat{g}+1}$. This proves the lemma.

To prove the theorem, we need only show that there is a limited set of d_i with the property that for any $m \in \mathbf{Z}^+$ and $(m, 2fd) = 1$, $(d_i/m) = 1 \Leftarrow mt^2 = F(x, y)$. In contrast to the statement in Lemma 8.2.16, the double implication is now reduced to the simpler direction!   □

8.3. Class transfer theory

We now consider the class field relation of K/k with $K/\mathbf{Q}$ normal and $k/\mathbf{Q}$ (cyclic) of prime order m. Thus,

(8.3.1a)
$$\left\{ \begin{array}{l} G_0 = \text{Gal } K/\mathbf{Q} = \{G + GT + \cdots + GT^{m-1}\} \\ \quad \text{(normal of order } nm) \\ G = \text{Gal } K/k \quad \text{(abelian of order } n) \\ g = G_0/G = \text{Gal } k/\mathbf{Q} = \langle t : t^m = 1 \rangle \\ \quad \text{(cyclic of prime order } m) \end{array} \right.$$

(8.3.1b)
$$\left\{ \begin{array}{l} K/k \leftarrow\text{WT}\rightarrow J \bmod^\times \tilde{\mathfrak{f}} \\ k/\mathbf{Q} \leftarrow\text{WT}\rightarrow j \bmod^\times \tilde{\mathfrak{f}}_0 \end{array} \right.$$

with the groups and classes designated by (abelian) G and H, where

$$G = \{S_1 (= 1), \ldots, S_n\} \leftarrow\text{A}\rightarrow J_{\uparrow}^*/J = H(J) = \{C_1 (= J), \ldots, C_n\}$$

and

$$g = \langle t : t^m = 1 \rangle \leftarrow\text{A}\rightarrow \tilde{\mathfrak{f}}_{f0}^*/j = H(j) = \langle c : c^m = j \rangle.$$

The relation $g = G_0/G$ for given G and g is possible for many non-isomorphic groups G_0 that contain G (i.e., an isomorphic image) as a subgroup. It does not necessarily follow, of course, that G_0 also contains g. This depends on whether or not G_0 splits at G. We shall see how the splitting depends on the transfer of classes from $H(j)$ to $H(J)$.

8.3.2. Definition. *The "transfer" of classes from $H(j)$ to $H(J)$ is a correspondence between each class c_s in $H(j)$ and a subset of classes $(C_{s_1}, \ldots, C_{s_t})$ in $H(J)$ determined by imbedding each ideal (a) in the class c_s in some class C_{s_i} that belongs to (a)O_k. [It is, of course, assumed that (a, $\mathfrak{f}\mathfrak{f}_0$) = 1.]*

8.3.3. Remark. It is clear that j must transfer into a subgroup of $H(J)$ and that other classes must transfer into whole cosets. This process forms only an injective image of the class group $H(j)$ in $H(J)$. [We recall that the ring class fields are determined by the property that *all* classes of $H(j)$ transfer into J.]

8.3.4. Lemma. *The principal class j transfers into a subgroup of $H(J)$ consisting of (say) r classes (C_{N_s}) that are norms of classes in $H(J)$:*

$$C_{N_s} = N_{k/\mathbf{Q}}C = C^{1+t+t^2+\cdots+t^{m-1}}, \qquad s = 1, \ldots, r.$$

8.3.4a. Lemma. *Equivalently, if $S_{N_s} \leftarrow A \rightarrow C_{N_s}$, then*

$$S_{N_s} = S^{1+T+T^2+\cdots+T^{m-1}}, \qquad S \leftarrow A \rightarrow C$$

for $T \in G_0$ while $T \notin G$ (T a coset generator of G_0/G).

Proof. These results follow from the splitting property (7.4.4a, b). □

To see how the nonprincipal class c transfers, first take a prime $(p) \in c$ (chosen, as usual, to remain unramified up to K, where its typical prime factor is denoted by $\mathfrak{P}$). Of course, (p) does not split from $\mathbf{Q}$ to k so its degree increases by a factor of m. Therefore, $K_Z^{\mathfrak{P}}$ does not contain k and $G_Z^{\mathfrak{P}}$ is not a subgroup of G (see Section 6.1). We can write the Frobenius generator of $G_Z^{\mathfrak{P}}$ as $[K/\mathbf{Q}/\mathfrak{p}] = T'$, where T' generates G_0/G. Then $T'G = T^uG$ ($\in g$) for some u, $(u, m) = 1$. We start by choosing generator c in such a way that, for simplicity, $u = 1$ and $T' = T$. So c transfers into $C_* \leftarrow A \rightarrow S_*$, where (by Lemma 6.4.3),

$$(8.3.4b) \qquad S_* = (K/k/(p)) = [K/k/\mathfrak{P}] = [K/\mathbf{Q}/\mathfrak{P}]^m = T^m.$$

If $u \neq 1$ or $T' \neq T$, we would replace S_* (and C_*) by $S_*^u S_{N_s}$ (and $C_*^u C_{N_s}$). This introduces the image of j from T^u/T' (see Lemma 8.3.4). We now obtain Lemma 8.3.5.

8.3.5. Lemma. *The nonprincipal classes c^u of $H(j)$, where $(u, m) = 1$, are associated to within a principal factor with C_*^u of equation (8.3.4b) so they transfer to the classes $(C_*^u C_{N_s})$, where $s = 1, \ldots, r$ and $(u, m) = 1$.*

In the quadratic case $m = 2$ ($u = 1$), the process is much simpler, as shown in the table (and discussed in Exercises 8.6–8.12). Note that the transfer of (p) is given together with that of c.

The special role of the quaternionic field is not so easily understood. It is best contrasted with the dihedral field. Let us call either type k_8 for now and note the tower of quadratic extensions:

$$(8.3.6) \quad \begin{cases} k_8 \supset k_4 \supset k_{2i} \supset \mathbf{Q} \\ k_{2i} = \mathbf{Q}(\sqrt{d_i}) \quad (d_1 d_2 d_3 = D^2) \\ k_4 = k_{21} \times k_{22} \times k_{23} = \mathbf{Q}(\sqrt{d_i}, \sqrt{d_{i+1}}), \quad i = 1, 2, 3. \end{cases}$$

(Note the cyclic numbering, $d_{i+3} = d_i$.)

In the quaternionic case, of course, there are no other subfields of k_8, and for each i,

$$\mathrm{Gal}\ k_8/k_{2i} = C(4).$$

In the dihedral case, there is a special quadratic field k_{21} with

$$(8.3.6a) \quad \begin{cases} \mathrm{Gal}\ k_8/k_{21} = C(4) \\ \mathrm{Gal}\ k_8/k_{22} = \mathrm{Gal}\ k_8/k_{23} = C(2) \times C(2). \end{cases}$$

(There are two other dihedral subfields of degree 4 between k_8 and k_{22} and also two between k_8 and k_{23}, which do not enter the discussion.) Quaternionic fields therefore have the higher degree of symmetry; all quadratic subfields are interchangeable under automorphisms of the Galois group.

8.3.7. Theorem (Witt). *For a given k_4, a quaternionic field k_8 exists if and only if the quadratic forms*

$$d_1 u_1^2 + d_2 u_2^2 + d_3 u_3^2 \quad and \quad u_1^2 + u_2^2 + u_3^2$$

Transfers for K cyclic of order 2, 3, 4 over k quadratic

$$G = C(2) \qquad (n = 2, \quad m = 2)$$

Four-group (dihedral of order 4), $G_0 = C(2) \times C(2) = D(2)$
$$G_0 = \langle S, T : ST = TS, \quad S^2 = T^2 = 1 \rangle$$
$$S^t = S, \quad S^{1+t} = 1, \text{ hence } j \to J$$
$$T^2 = 1, \quad (p) \to J, \text{ hence } c \to J$$

Cyclic group of order 4, $G_0 = C(4)$
$$G_0 = \langle S, T : S^2 = 1, \quad T^2 = S \rangle$$
$$S^t = S, \quad S^{1+t} = 1, \text{ hence } j \to J$$
$$T^2 = S, \quad (p) \to C, \text{ hence } c \to C$$

$$G = C(3) \qquad (n = 3, \quad m = 2)$$

Dihedral group of order 6, $G_0 = D(3)$
$$G_0 = \langle S, T : S^3 = 1, \quad T^2 = 1, \quad T^{-1}ST = S^{-1} \rangle$$
$$S^t = S^{-1}, \quad S^{1+t} = 1, \text{ hence, } j \to J$$
$$T^2 = 1, \quad (p) \to J, \text{ hence } c \to J$$

Cyclic group of order 6, $G_0 = C(6) = C(2) \times C(3)$
$$G_0 = \langle S, T : S^3 = 1, \quad T^2 = 1, \quad T^{-1}ST = S \rangle$$
$$S^t = S, \quad S^{1+t} = S^2, \text{ hence } j \to \{J, C, C^2\}$$
$$T^2 = 1, \quad (p) \to J, \text{ hence } c \to \{J, C, C^2\}$$

$$G = C(4) \qquad (n = 4, \quad m = 2)$$

Dihedral group of order 8, $G_0 = D(4)$
$$G_0 = \langle S, T : S^4 = T^2 = 1, \quad T^{-1}ST = S^{-1} \rangle$$
$$S^t = S^{-1}, \quad S^{1+t} = 1, \text{ hence } j \to J$$
$$T^2 = 1, \quad (p) \to J, \text{ hence } c \to J$$

Quaternionic group, $G_0 = Qu$
$$G_0 = \langle S, T : S^4 = T^4 = 1, \quad T^2 = S^2, \quad T^{-1}ST = S^{-1} \rangle$$
$$S^t = S^{-1}, \quad S^{1+t} = 1, \text{ hence } j \to J$$
$$T^2 = S^2, \quad (p) \to C^2, \text{ hence } c \to C^2$$

Noncyclic abelian group of order 8, $G_0 = C(2) \times C(4)$
$$G_0 = \langle S, T : S^4 = T^2 = 1, \quad T^{-1}ST = S \rangle$$
$$S^t = S, \quad S^{1+t} = S^2, \text{ hence } j \to \{J, C^2\}$$
$$T^2 = 1, \quad (p) \to J, \text{ hence } c \to \{J, C^2\}$$

Cyclic abelian group of order 8, $G_0 = C(8)$
$$G_0 = \langle S, T : S^4 = 1, \quad T^2 = S, \ T^{-1}ST = S \rangle$$
$$S^t = S, \quad S^{1+t} = S^2, \text{ hence } j \to \{J, C^2\}$$
$$T^2 = S, \quad (p) \to C, \text{ hence } c \to \{C, C^3\}$$

Note that only the dihedral groups G_0 belong to ring class fields.

are equivalent under unimodular rational transformations $[GL_3(\mathbf{Q})]$. *In terms of the Hilbert symbols (see the exercises in Chapter 2),*

(8.3.7a) $(d_1, d_2/p) = (-1, d_1 d_2/p)$

over the finite and infinite primes. (Thus, at the very least, all k_{2i} *must be real.)*

8.3.8. Theorem. *The quaternionic extensions* k_8/k_{2i} *can never be unramified.*

Proof. If k_8/k_{2i} is unramified, the conductor $\mathfrak{f} = 1$ so all rational ideals must behave the same for k_{2i} to k_8 (merely split). This contradicts the transfer in the table (of j into J and c into C^2).   □

The construction of quaternionic fields is rather complicated, and we merely refer to work in the bibliographic note at the end of the chapter. We return to the structure (8.3.6) for the next result.

8.3.9. Theorem (Redei and Reichardt). *For a given* k_4, *we have an unramified* k_8 *(dihedral) if* $d_1 = d_2 d_3$ *and the equation*

(8.3.9a) $x_1^2 = d_2 x_2^2 + d_3 x_3^2$

is solvable in **Z**. *Then,*

(8.3.9b) $k_8 = k_{21}(\sqrt{(x_1 + x_2\sqrt{d_2})})$.

Indeed if d_1 *or* $d_1/8$ *is odd, these are the only possible* k_8. *(The factorization* $d_1 = d_2 d_3$ *is called a "Redei decomposition" and several such may exist for any* d_1.)

Actually, the transfer process for a quaternionic extension is more easily seen using all four classes associated with

$$(8.3.10) \begin{cases} k_4/\mathbf{Q} \leftarrow\text{WT}\rightarrow j_* \\ j_D^*/j_* = \{j_*, c_1, c_2, c_3\}, \quad (D \text{ a suitable conductor}) \\ j_* = \{(a) : (d_i/a) = 1, \text{ all } i\} \\ c_i = \{(a) : (d_i/a) = 1, (d_k/a) = -1 \quad \text{for } k \neq i\}. \end{cases}$$

Then $k_{2i}/\mathbf{Q} \leftarrow\text{WT}\rightarrow \{j_*, c_i\}$ (the residues of d_i). In the transfer of $k_{2i}/\mathbf{Q}$ to k_8/k_{2i},

$$\{j, c_i\} \to J_i(\leftarrow \text{WT} \to k_8/k_{2i})$$
$$\{c_{i+1}, c_{i+2}\} \to C_i^2$$

where Gal $k_8/k_{2i} = H(J_i) = \langle C_i : C_i^4 = J_i \rangle$. {If we were dealing with a dihedral extension, then for some special quadratic subfield k_{2i} [see equations (8.3.6a)], all classes c_i and j_* would go into J_1 on this transfer.}

We now interpret this process in terms of "good" versus "bad" principalization in Hilbert class fields, as in Illustration 7.1.6. To do so we shall start not with **Q** as in the base field, but with

$$k_{2*} = \mathbf{Q}(\sqrt{d_1 d_2 d_3})$$

and note the genus field

$$k_{8*} = \mathbf{Q}(\sqrt{d_1}, \sqrt{d_2}, \sqrt{d_3}).$$

If $h = 4$, then k_{8*} is the Hilbert class field; that is,

$$\text{Gal } k_{8*}/k_{2*} = C(2) \times C(2) = H(k_{2*}).$$

As before, $k_{8*}/k_{2*} \leftarrow \text{WT} \to j_*$ and now j_* is principal whereas c_i are the nonprincipal classes. Suppose further that k_{8*} is of class number 2 so it has a Hilbert class field k_{16}. Since k_{16}/k_{2*} is normal and nonabelian (k_{8*} is maximal unramified abelian), there is the choice: Is k_{16}/k_{2*} dihedral or quaternionic?

8.3.11. Theorem (Taussky and Kisilevsky). *In the quaternionic case, we have good principalization (i.e., the class c_i becomes principal only in k_{2i}), whereas in the dihedral case, we have bad principalization (i.e., all classes become principal in all k_{2i}).*

For a complete proof, we refer to the references, but it may be noted that the part involving the quaternionic field follows from the earlier transfer argument. A class that transfers into c_i^2 does not become principal until k_{8*}.

8.3.12. Theorem. *Let $k_1 = \mathbf{Q}(\sqrt{d_1}) \neq k_2 = \mathbf{Q}(\sqrt{d_2})$ and let K be a ring class field over k_1 and k_2. Then K is of the composite quadratic type: $K = \mathbf{Q}(\sqrt{m_1}, \ldots, \sqrt{m_r}) \supseteq k_1 \times k_2 = k_{12}$.*

Proof. Let $G = \text{Gal } K/\mathbf{Q}$ and let the various subfields be fixed by $G_1 \leftarrow\text{Gal}\rightarrow k_1$, $G_2 \leftarrow\text{Gal}\rightarrow k_2$, and $G_{12} \leftarrow\text{Gal}\rightarrow k_{12}$. Then G_{12} is a subgroup of index 2 in both G_1 and G_2, all of which are abelian groups. We must show that G is abelian of type $C(2) \times \cdots \times C(2)$.

We can construct G_1, G_2, G_{12} by using the Hilbert sequence. Let $t = 1$ or 2 for the fields k_t (so $t \pm 1$ denotes the other one). Then we define p_t as a nonramifying prime such that

$$(d_t/p_{t\pm 1}) = 1, \qquad (d_t/p_t) = -1.$$

So $p_{t\pm 1}$ is inert from $\mathbf{Q}$ to k_t and therefore splits from k_t to K. Therefore, if $\mathfrak{P}_t \mid p_t$ in K, then $\mathfrak{P}_t$ is of order 2 as is the Hilbert splitting group of $\mathfrak{P}_t$, namely, $\langle T_t : T_t^2 = 1 \rangle$. Now T_t must lie in G_t and not in $G_{t\pm 1}$ because it fixes a field that contains k_t and not $k_{t\pm 1}$. As a result,

$$G_1 = G_{12} + T_1 G_{12}, \qquad T_1 \notin G_{12}, \qquad T_1 \notin G_2$$
$$T_1^2 = 1, \qquad T_1\sqrt{d_1} = \sqrt{d_1}, \qquad T_1\sqrt{d_2} = -\sqrt{d_2},$$
$$G_2 = G_{12} + T_2 G_{12}, \qquad T_2 \notin G_{12}, \qquad T_2 \notin G_1$$
$$T_2^2 = 1, \qquad T_2\sqrt{d_2} = \sqrt{d_2}, \qquad T_2\sqrt{d_1} = -\sqrt{d_1}.$$

We noted that G_1, G_2, and $G_{12} = \{U, V, \ldots\}$ are all abelian. Therefore, if we decompose

$$G = G_{12} + T_1 G_{12} + T_2 G_{12} + T_1 T_2 G_{12},$$
$$T_1 U T_1^{-1} = U, \qquad T_2 U T_2^{-1} = U.$$

On the other hand,

$$T_2 S_1 T_2^{-1} = S_1^{-1}, \qquad T_1 S_2 T_1^{-1} = S_2^{-1}$$

for $S_i \in G_i$, by the property that $C' = C^{-1}$ (Lemma 8.2.11). Putting these two results together, we find

$$T_2 U T_2^{-1} = U^{-1} = U, \qquad U^2 = 1$$

whereas $T_2(T_1 T_2^{-1} T_1^{-1}) = T_2^2 = T_1^{-2} = 1$. So by these relations all elements X of G, that is, U, $T_t U$, and $T_{12} U$, satisfy $X^2 = 1$. Hence G is abelian (of the desired type), so K is composite quadratic. $\qquad\square$

Exercises

The following exercises illustrate the use of quadratic forms of discriminant df^2 as an isomorphic model of proper ideals of the integer ring O_f. These are, in turn, isomorphic to ideal classes in ring class fields of conductor f over the quadratic field.

8.1. Verify the following correspondences of ideals and forms:
a. $d = -4 \times 14$, $h = h_+ = 4$ (cyclic).

$$[1, \sqrt{-14}] \leftrightarrow x^2 + 14y^2, \qquad [2, \sqrt{-14}] \leftrightarrow 2x^2 + 7y^2$$
$$[3, 1 + \sqrt{-14}] \leftrightarrow 3x^2 + 2xy + 5y^2,$$
$$[3, -1 + \sqrt{-14}] \leftrightarrow 3x^2 - 2xy + 5y^2.$$

b. $d = 4 \times 79$, $h = 3$, $h_+ = 6$.

$$[1, \sqrt{79}] \leftrightarrow x^2 - 79y^2, \qquad \sqrt{79}[1, -\sqrt{79}] \leftrightarrow -x^2 + 79y^2$$
$$[3, 1 + \sqrt{79}] \leftrightarrow 3x^2 + 2xy - 26y^2,$$
$$\sqrt{79}[3, -1 + \sqrt{79}] \leftrightarrow -3x^2 - 2xy + 26y^2$$
$$[3, -1 + \sqrt{79}] \leftrightarrow 3x^2 - 2xy - 26y^2,$$
$$\sqrt{79}[3, -1 - \sqrt{79}] \leftrightarrow -3x^2 + 2xy + 26y^2.$$

8.2. Consider O_4 for $\mathbf{Q}(i)$. Show that $h(-64) = 2$ and that these two classes are represented by (1) and $(2 + i)$ (no longer principal!).

$$(1) = [1, i], \qquad [1, i] \cap O_4 = [1, 4i] \leftrightarrow x^2 + 16y^2$$
$$(2 + i) = [5, 2 + i], \qquad [5, 2 + i] \cap O_4 = [5, 3 + 4i]$$
$$\leftrightarrow 5x^2 + 6xy + 5y^2.$$

8.3. For $\mathbf{Q}(\sqrt{-3})$ when $f = 6$, then $h = 3$. Show that the corresponding forms are $x^2 + 27y^2$ and $7x^2 \pm 2xy + 4y^2$. When $f = 9$, $h = 3$, again leading to three forms, $x^2 + xy + 61y^2$ and $7x^2 \pm 3xy + 9y^2$. Explain the relation to ring class fields over $k = \mathbf{Q}(\sqrt{-3})$: Here $k(\sqrt[3]{2})/k \leftarrow\text{WT}\rightarrow J(2)$ and $k(\sqrt[3]{3})/k \leftarrow\text{WT}\rightarrow J(3)$. Hence, $p = x^2 + 27y^2 \leftrightarrow w^3 \equiv 2 \bmod p$ (solvable for w) and $p = x^2 + xy + 61y^2 \leftrightarrow w^3 \equiv 3 \bmod p$ (solvable for w). (Always $p \equiv 1 \bmod 3$ splits in $k/\mathbf{Q}$.)

8.4. Show that analogously with Theorem 8.1.13 there is a weak genus theory applicable to the case where k is real but K_g is complex. Thus, $d = d_0 \cdots d_r$ where (say) $d_0 < 0$. Then K_g^0, the weak genus field, is the maximal real field in K_g. It is generated by $\sqrt{d_r}$ if $d_r > 0$ and by $\sqrt{(d_0 d_r)}$ if $d_r < 0$. It is the class field that corresponds to the ideal group $J = \{a^2 J_1'\}$ (the ideal classes weakly equivalent to squares).

8.5. The concept of "composition" refers to the classical identity $(x_1^2 + y_1^2)(x_2^2 + y_2^2) = (x_3^2 + y_3^2)$ based on norms in the product $(x_1 + iy_1)(x_2 + iy_2) = (x_3 + iy_3)$. Here there is a bilinear relation between (x_3, y_3) and (x_1, y_1), (x_2, y_2). Correspondingly, if we write $(ax_1 + (b + \sqrt{d})y_1/2)(a_0 x_2 + (b + \sqrt{d})y_2/2) = (aa_0 x_3 + (b + \sqrt{d})y_3/2)$, then, on expanding, we obtain a bilinear relation that composes the forms $ax_1^2 + bx_1 y_1 + ca_0 y_1^2$ and $a_0 x_2^2 + bx_2 y_2 + cay_2^2$ to form $aa_0 x_3^2 + bx_3 y_3 + cy_3^2$. Show how this is parallel to the multiplication of the modules $[a, (b + \sqrt{d})/2]$ and $[a_0, (b + \sqrt{d})/2]$.

Consider $G_0/G = g$, with G and g given from now on by

$$G = C(n) = \langle S : S^n = 1 \rangle, \qquad G_0 = G + GT$$
$$g = C(2) = \{G, GT\} \qquad (T \notin G).$$

For each n we define G_0 by integers a and b modulo n:

$$G_0 = \langle S, T : S^n = 1, \quad T^2 = S^a, \quad T^{-1}ST = S^b \rangle.$$

8.6. Note that S and T are not unique. We can replace S by S^v where $(v, n) = 1$ and we can replace T by TS^u. Under such changes, verify that b is still well defined mod n, whereas a is defined only mod A where $(0 <) A = (n, b + 1)$. In fact, a can be replaced by any member of the subset $\{av\}$ mod A where $(v, n) = 1$. (We agree to choose a to be the minimal such value.)

8.7. Verify that G_0 is of degree $2n$ (as required) exactly when $b^2 \equiv 1$ and $a(b - 1) \equiv 0$ mod n.

8.8. Verify that the transfers are as follows:

$$j \to \{C^r : r \equiv 0 \bmod A\}$$
$$c \to \{C^s : s \equiv av \bmod A, (v, n) = 1\} \quad \text{where, as usual, } C \leftarrow A \to S.$$

8.9. Verify that these groups occur for all n:
a. Dihedral $D(n)$; $a = 0, b = -1, A = n$:

$$G_0 = \langle S, T : S^n = 1, \quad T^2 = 1, \quad T^{-1}ST = S^{-1} \rangle$$
$$j \to J, \quad c \to J \text{ (ring class field)}.$$

b. Cyclic $C(2n)$; $a = 1, b = 1, A \equiv n \bmod 2$:

$$G_0 = \langle S, T : S^n = 1, \quad T^2 = S, \quad T^{-1}ST = S \rangle$$
$$n \text{ odd}, \quad A = 1, \quad j \to \{C^r, \text{ all r}\}, \quad c \to \{C^s, \text{ all s}\}$$
$$n \text{ even}, \quad A = 2, \quad j \to \{C^r, \text{ even r}\}, \quad c \to \{C^s, \text{ odd s}\}.$$

Verify that these groups occur, in addition, for even n:
c. Dicyclic $S(n)$; $a = n/2, b = -1, A = n$:

$$G_0 = \langle S, T : S^n = 1, \quad T^2 = S^{n/2}, \quad T^{-1}ST = S^{-1} \rangle$$
$$j \to J, \quad c \to C^{n/2} \text{ } (n = 4 \text{ is quaternionic}).$$

d. Bicyclic $C(n) \times C(2)$; $a = 0, b = 1, A = 2$:

$$G_0 = \langle S, T : S^n = 1, \quad T^2 = 1, \quad T^{-1}ST = S \rangle$$
$$j \to \{C^r, \text{ even r}\}, \quad c \to \{C^s, \text{ even s}\}.$$

Verify that these are the only groups for $n = 2, 3, 4$.

8.10. When $n = 8$, $b^2 \equiv 1 \bmod n$ has many solutions. They lead to two groups in addition to those listed in Exercise 8.9, corresponding to $a = 0$, $b = \pm 3$. Find the transfers.

8.11. Verify that only the dihedral case is a ring class field. *Hint:* In Exercise 8.8, $c \to 1$ (only) exactly when $A = n$ and $a \equiv 0 \bmod n$.

8.12. When n is prime or twice prime, the transfers uniquely determine the group G_0. Is this the only such case? *Hint:* Consider Exercise 8.10.

8.13. Verify the following four instances of Exercises 8.9a–d:

a. $K = \mathbf{Q}(\sqrt{-7}, \sqrt{2}, \sqrt{-1 + 2\sqrt{2}})$, $k = \mathbf{Q}(\sqrt{-14})$, Gal $K/k = D(4)$.

[This is the Hilbert class field for k, which has class group $C(4)$.] Show that K/k is unramified, and j and c [the residues and nonresidues of $(-14/\ldots)$] both transfer into J, the principal ideal class of k.

b. $K = \mathbf{Q}(\sqrt{(2 + \sqrt{(2 + \sqrt{2})})})$, $k = \mathbf{Q}(\sqrt{2})$, Gal $K/k = C(8)$.

This is the real subfield of $\mathbf{Q}(\exp 2\pi i/32)$. Verify that if $j = (m)$, then j transfers into J if $m \equiv \pm 1 \bmod 16$, otherwise into C^2.

c. $K = \mathbf{Q}(\sqrt{(2 + \sqrt{2})(3 + \sqrt{3})})$, $k = \mathbf{Q}(\sqrt{2})$, Gal $K/k = Qu$.

Here K has no subfields of degree 4 except $\mathbf{Q}(\sqrt{2}, \sqrt{3})$ and hence is quaternionic. (This is a remarkable exercise in the use of units!) Check the transfers.

d. $K = \mathbf{Q}(\sqrt{(2 + \sqrt{2})}, \sqrt{3})$, $k = \mathbf{Q}(\sqrt{3})$, Gal $K/k = C(2) \times C(4)$.
Check the transfers.

Bibliographic note

For ring class field theory, see Bruckner (1966), Cohn (1978), Deuring (1958), Fricke (1928), and (for a generalization) Satge (1977). For genus theory, also see Cohn (1979b), Fröhlich (1959, 1983), Ishida (1976), Jones (1950), and Shanks (1971). Transfer theory is best described in Hasse (1949); also see Chebotarev (1930) and Cohn (1978). Additional relative quadratic applications are numerous, for example, Scholz (1934), Rédei and Reichardt (1934), Cohn and Barrucand (1969), Witt (1936), and Kisilevsky (1976).

9

Class fields by radicals

Because class fields over k are just the abelian extensions over k, we could presumably construct class fields by constructing sufficiently general types of abelian fields over k and selecting those with whatever property is required. Thus, for a ray class field given by a factorization property, we might specify a conductor or a discriminant. For a Hilbert class field given by a principalization property, we might specify the ideal class or group that is to be principalized. In either case, there are occasions where it is of considerable help to go back to the idea that abelian extensions are constructible by radicals, so-called Kummer extensions $K = k(\mu^{1/l})$.

As an illustration, $K = k(\sqrt{5})$ is relatively abelian and unramified over $k = \mathbf{Q}(\sqrt{-5})$. Here k has class number two, with $2 = 2_1^2$, where $2_1 = (2, 1 + \sqrt{-5})$, a generator of the nonprincipal class. We can see K/k as a ring class field of unit relative discriminant, since $(5) = (\sqrt{-5})^2$ in k and $5 \equiv 1 \bmod 4$ (see Lemma 6.2.18). We can also see K/k as a Hilbert class field in which 2_1 becomes principal. Note that $K = k(i)$ as well; therefore, $(1 + i)$ and 2_1 each has (2) as a square in K and thus $2_1 = (1 + i)$ in K. In either treatment the adjoined radical describes the field as a class field.

In this chapter we generalize on this example (as did Hilbert). We take only a scattered sample of problems, but we hope to display the interplay of principalization, factorization, and algebra with hard number theory, particularly the theory of units.

9.1. An unramified class field tower

One of the more challenging motivations for the construction of class fields is the principalization of ideals. We can well assume Theorem

7.4.17, which states that all ideals principalize in the Hilbert class field, but we must still address the problem that the ideals might principalize in smaller fields (seemingly unpredictably).

9.1.1. Theorem (Hilbert's 94). *Let K/k be a cyclic extension of prime degree l unramified over k (also at ∞). Then there exists (at least) one class of nonprincipal ideals $\mathfrak{a}$ in k that is of order l and that becomes principal on injection into K [i.e., $\mathfrak{a}O_K = (A)$].*

This result has the advantage that its proof is constructional and elementary (see Lemma 9.2.4). The motivation for the proof can be seen if we consider a special case (see Lemma 6.2.17). Let

$$(9.1.2) \quad K = k(\sqrt[l]{\mu}), \qquad \mu \equiv \alpha^l \bmod (1 - \zeta)^l, \, \alpha \in o, \qquad (\alpha, l) = 1$$

where k contains $\zeta = \exp 2\pi i/l$. Here K/k is unramified if $(\mu) = \mathfrak{a}^l$ for some ideal $\mathfrak{a}$. Assume $\mathfrak{a}$ is not principal. We can now see that $\mathfrak{a}$ is of order l but that it principalizes; in fact, it becomes $\mu^{1/l}$ as we go from k to K. The early attempts to prove that all ideals principalize in a suitable extension field were based on such examples, but more powerful algebraic methods must be brought to bear on the problem. (For Chapter 9, we do not assume Theorem 7.4.17 on principalization in the Hilbert class field, but we do occasionally need the WT-correspondence in Theorem 7.4.1.)

9.1.3. Theorem. *Let $Cl\{k\} = C(l^r) \times G$ where $|G|$ is prime to l. Then there is a tower of fields, an "l-tower,"*

$$(9.1.3a) \quad K = k_r \supset k_{r-1} \supset \cdots \supset k_1 \supset k_0 = k, \qquad [k_i : k_{i-1}] = l$$

and an ideal $\mathfrak{a}$ in k of order l^r such that in each field k_i the injection $\mathfrak{a} \to \mathfrak{a}O_{k_i} = \mathfrak{a}_i$ is an ideal of order l^{r-i}. In particular, $\mathfrak{a}_r = \mathfrak{a}O_K$ is principal.

9.1.4. Definition. *We call an ideal $\mathfrak{a}$ of k "l-principal" when the order of (the strict class of) $\mathfrak{a}$ is prime to l. Such ideals are a group, and there is an "l-class group, namely $Cl_l\{k\}$, defined relative to it. Likewise there is an "l-Hilbert class field" (of unit conductor $K = H_l(k)$ defined by WT-correspondence with $Cl_l\{k\}$ (with the property that $Gal \, K/k$ is the l-class group). As an alternate definition, K is the maximal unramified abelian extension of k for which the l-principal prime ideals split.*

To prove Theorem 9.1.3 assume that $Cl_l\{k\} = C(l^r)$ (cyclic) and that $K = H_l(k)$. It will have the subfields over k shown in equation (9.1.3a) (and none else). We first need the following lemma.

9.1.5. Lemma. *For each k_i in the tower (9.1.3a) (including K),*

$$H_l(k_i) = H_l(k) \quad (= K).$$

(Thus wherever we start, the l-class field tower goes precisely up to K.)

Proof. We note that $L = H_l(k_i)$ contains $K = H_l(k)$, since by definition L is the maximal relatively abelian field over k_i in which all l-principal primes split. In fact, under the norm mapping of ideals from k_i into k, namely

(9.1.5a)
$$N_i : \mathfrak{a} \to N_{k_i/k}\mathfrak{a}$$

the primes from k_i map into those of k, preserving splitting under the map. Now if we set $G = \text{Gal } L/k$ and consider the tower

$$L \supseteq K \supseteq k_i \supseteq k$$

then by the Galois correspondence (for G' the commutator),

$$L \leftrightarrow 1, \quad k \leftrightarrow G', \quad K \leftrightarrow G.$$

This is true because K is the maximal abelian unramified extension of k below L. Now $\text{Gal } K/k = G/G' = C(l')$, so we conclude that $K = L$ from the next purely group-theoretic result $\qquad\qquad\square$

9.1.6. Theorem. *If G is an l-group (group whose order is a power of l) and if G/G' is cyclic, then $G' = 1$ and G itself is cyclic.*

The proof is an immediate consequence of the Burnside basis construction (given here in the context of an arbitrary l-group G). All (proper) maximal subgroups G_i of G are known to be normal with G/G_i (abelian) of order l. This informs us that each G_i contains all l-powers and also the commutator subgroup G'; so does the intersection of all G_i, which we call D. Accordingly, in coset form,

$$G/D = \langle a_1 D, \dots, a_r D \rangle = C(l)^r \quad (a_i \in G).$$

Now the group $H = \langle a_1, \dots, a_r \rangle$, generated by the so-called Burnside basis, is actually all of G. If, to the contrary, H (and HD) are contained in G_0, a maximal subgroup in G, then $|G/D| > |G_0/D| = l^{r-1}$, which contradicts $|HD/D| = l^r$. In our special context G/G' ($\supseteq G/D$) is cyclic, so $r = 1$ and, accordingly, G is cyclic.

To continue the proof of Theorem 9.1.3, note that now the splitting primes from k_i to K are seen to be mapped onto the same set (those in

k) for each mapping N_i in expression (9.1.5a), since K is the common class field. In particular,

$$(9.1.7a) \qquad \text{Cl}_l\{k_i\} = C(l^{r-i}) = H_i(k_i) = \text{Gal } K/k_i$$

so $\text{Cl}_l\{K\} = 1$ and all ideals of k and l-principal in K.

We consider another mapping, the injective mapping from k to k_i,

$$(9.1.7b) \qquad J_i : \mathfrak{a} \to \mathfrak{a}O_{k_i}.$$

We now assert that under J_i the order of a generating ideal class is divided by l^i. For proof let $\mathfrak{g}$ be an ideal in k generating its l-class group, so $\mathfrak{g}$ has order l. Then $\mathfrak{g}_i = J_i\mathfrak{g}$ is of order l^-_i (times factors prime to l) and it generates $\text{Cl}_l\{k_i\}$. This is true because $N_i\mathfrak{g}_i = \mathfrak{g}^{l^i}$ (all conjugates over k are the same). So if $\mathfrak{g}_i$ were of any lower order in k_i, $\mathfrak{g}$ would also be lower in k.

This completes the proof and provides us with a perfect paradigm of good principalization (see Section 7.1) as follows: For the tower (9.1.3a), we make the correspondences over k given by

$$(9.1.7c) \qquad k_i/k \leftarrow \text{WT} \to G_i = \langle g^{l^{r-1}} : g^{l^r} = 1 \rangle \qquad (|G_i| = l^i)$$

so $|G_i|$ accounts for the ideals that are l-principal in k_i. (If we raise the ideal classes to some power prime to l, we can make all these ideals actually principalize.) In any case, there is no "premature or unpredictable" principalization through taking l-towers.

These methods were communicated by Olga Taussky, who applied and extended them to difficult cases where the class group is not cyclic. We now consider the actual construction of the fields in the tower by the adjunction of radicals. We give a result of Cooke.

9.1.8. Theorem. *If k contains a primitive l-root of unity, then in the l-tower (9.1.3a),*

$$(9.1.8a) \qquad k_{i+1} = k_i(\mu_i)^{1/l}$$

where as ideals in $k_i, (\mu_i) = \mathfrak{a}_i^l$ [written simply as (1) if μ_i is a unit]. The l-factor in the order of the class of $\mathfrak{a}_i$ is independent of the choice of the μ_i that generates k_i. Thus, in O_{k_i} either $\mathfrak{a}_i$ is of order l (so $\mathfrak{a}_i^l$ is l-principal) or else we can take μ_i as a unit (if $\mathfrak{a}_i$ is l-principal).

If, furthermore, k contains a primitive l^2-root of unity, then μ_i can always be taken as a unit. Indeed, then k_{i+2}/k_i is achieved by a radical as

$$(9.1.8b) \qquad k_{i+2} = k_i(\omega_i)^{1/l^2}$$

but $(\omega_i) = \mathfrak{b}_i^{l^2}$ *for some ideal* $\mathfrak{b}_i$ *of order either 1 or l (referring, of course, to l-factors of class orders in k_i).*

Proof. The first part of the result follows from the fact that if μ_i is one radicand for k_{i+1}/k_i, then the most general radicand is $\mu_i^a \rho^l$ for $\rho \in k$ and $(a, l) = 1$. (Compare Remark 6.2.16 and Lemma 6.2.17.) The second part combines equations (9.1.8a) and (9.1.8b) in two different ways. We write

(9.1.8c) $$k_{i+1} = k_i(\mu_i)^{1/l} = k_i(\omega_i)^{1/l}$$

so by comparison, in the notation of equations (9.1.8a, b),

(9.1.8d) $$\mathfrak{a}_i \sim \mathfrak{b}_i^l \times (l\text{-factors})$$

(i.e., both ideals have the same order except for factors prime to l). Furthermore,

(9.1.8e) $$k_{i+2} = k_{i+1}(\mu_{i+1})^{1/l} = k_{i+1}((\omega_i)^{1/l})^{1/l}$$
(9.1.8f) $$\mathfrak{a}_{i+1} \sim \mathfrak{b}_i O_{k_{i+1}} \times (l\text{-factors}).$$

The injection of $\mathfrak{b}_i$ into k_{i+1} here produces exponentiation by l for the order of each class in the ideal group, just as in expression (9.1.8d). It is now clear that in the field k_i, the ideal $\mathfrak{a}_i$ has the same order in k_i as $\mathfrak{a}_{i+1}$ has in k_{i+1}; this is possible only when both orders are 1. Remarks on the order of the ideals $\mathfrak{b}_i$ follow.   □

9.1.9. Illustration of the two-class tower for $Q(\sqrt{-p})$. As an application, consider the field $k = Q(\sqrt{-p})$, which by genus theory has a two-class structure of $C(2^r)$. We assume $p \equiv 1 \pmod 8$ or $r \geq 2$. Here

(9.1.9a) $$k_1 = k(i)$$

so Theorem 9.1.8 will apply from this point on (with $l = 2$). To construct k_2, we need to decide whether the radicand is a unit. This is answered by

(9.1.9b) $$k_2 = k_1(\epsilon)^{1/2} \qquad (\epsilon > 0)$$

for ϵ the fundamental unit of $Q(\sqrt{p})$. It is an elementary fact that $\epsilon = a + b\sqrt{p}$, where a and b are integers and $4 \mid a$, if we consider the equation $-1 = a^2 - pb^2 \bmod 8$. Thus,

$$[(\epsilon - 1)/2][(\epsilon' - 1)/2] = a/2 \equiv 0 \bmod 2.$$

But in k_1 (of degree 4), $2 = 2_1^2 2_2^2$, whereas $(\epsilon - 1)/2$ and $(\epsilon' - 1)/2$ are relatively prime. So one of these is divisible by 2_1^2 and the other is divisible by 2_2^2. Thus,

$$\epsilon \equiv 1 \bmod 2_1^4, \qquad \epsilon' \equiv 1 \bmod 2_2^4.$$

Because $\epsilon = -1/\epsilon'$, $\epsilon \equiv i^2 \bmod 2_2^4$ by the Chinese remainder theorem. We seek an integer γ such that

$$\gamma \equiv 1 \bmod 2_1^4, \qquad \gamma \equiv i \bmod 2_2^4.$$

Then $\epsilon \equiv \gamma^2 \bmod 4$, which justifies equation (9.1.9b). (Each step in the tower is uniquely determined by nonramification.) To go higher in the tower we set up nonramified fourth-degree radicals. The technical details are given in the papers of Cohn and Cooke (see Bibliographic Note). The highest field in the tower is $RCF_2\{-4p\}$, the 2-Sylow ring class field (see Section 2.2).

We give the next step when $r \geq 3$. We define the parameters

$$p = 2e^2 - f^2, \qquad f \equiv -1 \bmod 4.$$

Then to generate the next step by a fourth-degree radical, take

(9.1.9c) $k_3 = k_1(\omega_1)^{1/4}, \qquad \omega_1 = \epsilon(f + \sqrt{-p})^2/(1 - i)^2.$

Of course, k_3 is also $k_2(\omega_1)^{1/4}$, but it can be expressed as well in terms of quadratic radicals of units (see Section 9.2).

The same field k_3 can also be generated by a unit; for example,

(9.1.9d) $$k_3 = k_2(\sqrt{(\sqrt{\epsilon \theta_2 \theta_2^c})})$$

where θ_2 is indeed a unit of the form

(9.1.9e) $\theta_2 = [(a + bi) + \sqrt{\pi}(c + di)]/2, \qquad \pi = u + iv$
$\pi\pi' = u^2 + v^2 = p.$

Here π, π' and θ, θ^c represent complex conjugate pairs; $a + bi$ and $c + di$ are Gaussian integers that remain to be chosen so that

(9.1.9f) $$\theta_2 \theta_2' = [(a + bi)^2 - \pi(c + di)^2]/4 = i$$

satisfying a Pellian equation (with θ' denoting the algebraic conjugate where the sign of $\sqrt{\pi}$ is changed). Actually, θ_2 belongs to k_2; thus,

(9.1.9g) $$k_3 = k_2(\sqrt{U_2})$$

for U_2 a unit of k_2.

9.2. A ramified tower

A ramified tower can serve a purpose to supplement that of the unramified tower (as seen in Section 9.1). We recall that an unramified tower $k \subset K \subset \cdots$ displays decreasing class numbers (in some l-adic sense). A ramified tower can keep class numbers from growing (in the same sense) if the ramification is the "least possible."

In this section we construct ramified Kummer extensions using a condensed version of Hilbert's theory. A complete version (as carried out by Furtwängler and others subsequently) would in essense approximate the WHAT theorem. This abridgement will, for the time being, not assume the WHAT theorem but is intended to illustrate useful techniques, particularly involving units, for the construction of class fields.

We let $K = k(\sqrt[l]{\mu})$, where k contains ζ, a primitive lth root of unity (l prime). Let k be of degree $m = r_1 + 2r_2$ with $r = r_1 + r_2 - 1$, the (torsion-free) rank of the units o^* of k. If l is odd, of course $m = 2r_2$, but if $l = 2$, we make the further assumption that for the r_1 real conjugates of k, the field $K = k(\sqrt[l]{\mu})$ is also real. (In other words, we are assuming that no ramifications occur at ∞.) Thus, the real quadratic fields over $\mathbf{Q}$ are included but not the imaginary ones. The symbol N will always mean relative norm (i.e., $N_{K/k}$) and

$$\text{Gal } K/k = \langle S : S^l = 1 \rangle.$$

Thus, if $A \in K$,

$$(9.2.1) \quad NA = AA^S \cdots A^{S^{l-1}} = A^{1+S+\cdots+S^{l-1}} \quad (A^{S^l} = A).$$

Of course, the elements of k (as well as the ideals in k, etc.) are invariant under S.

We begin with an elementary result:

9.2.2. Hilbert's Theorem 90. *If $A \in K$ and $NA = 1$, then $A = B^{1-S}$ for B an integer in k.*

The converse is trivial as $(1 - S)(1 + S + S^2 + \cdots + S^{l-1}) = 1 - S^l$ and we shall note that the theorem holds even for any cyclic extension (Kummerian or otherwise) of any degree (prime or otherwise).

Proof. Let x be an as yet unknown integer in $\mathbf{Z}$. Define

$$A_x = (1 + x\theta)A/(1 + x\theta^S)$$

for θ a generator of k/k. Thus, $NA_x = NA = 1$. Likewise, define the rational function of x:

$$(9.2.2a) \qquad B_x = 1 + A_x + A_x^{1+S} + \cdots + A_x^{1+S+\cdots+S^{l-2}} = A_x B_x^S$$

so $B_x^{1-S} = A_x$. Now B_x does not vanish identically in x, since it would do so with x a complex parameter as well, whereas it has an obvious pole when $1 + x\theta^{S^i} = 0$ ($i > 0$). Hence we can find a rational integer a such that $B_a \neq 0$. Finally, we define $B\ (\in K)$ by

$$(9.2.2b) \qquad\qquad B = MB_a/(1 + a\theta)$$

and for this B, $B^{1-S} = A$, where M is a rational integer chosen so as to make B an integer. Note that for $l = 2$, we customarily choose $a = 0$ so $B = M(1 + A)$, which leads to the simple identity $A = (1 + A)/(1 + A^S)$, a consequence of $NA = AA^S = 1$. [There is, of course, the exceptional case $A = -1$, where an $x \neq 0$ could lead to $B = \sqrt{D}$ if $K = k(\sqrt{D})$.] $\qquad\square$

9.2.3. Hilbert's Theorem 92.

For K/k cyclic of prime degree, there always exists a unit $H \in K$ such that $NH = 1$ but $H \neq E^{1-S}$ for any $E \in O^*$.

The proof is deferred until item 9.3.4. We note that as a consequence $H = A^{1-S}$ for $A \notin O^*$, so that $(A) = \mathfrak{A}$, a principal ideal in O. Thus, $\mathfrak{A}^S = (A^S) = (A/H) = \mathfrak{A}$. Therefore, if $NA = \alpha$ (in o), then

$$(9.2.3a) \qquad N\mathfrak{A} = \mathfrak{A}\mathfrak{A}^S \cdots \mathfrak{A}^{S^{l-1}} = \mathfrak{A}^l = (NA) = \alpha O.$$

9.2.4. Lemma.

In the preceding notation, if $\mathfrak{A}$ lies in o [i.e., $(A) = \mathfrak{A} = aO$ for some ideal a in o], then a is nonprincipal in o whereas a^l is principal. (Thus, the class number of k is divisible by l and the class of the ideal a principalizes from k to K, possibly together with other classes that are not generated by a.)

Proof. If $\mathfrak{A}$ lies in o, then a must be nonprincipal. Otherwise, $\mathfrak{A} = E\beta$ for $E \in O^*$, and $(\beta) = a$ with $\beta \in o$. This would lead to $H = E^{1-S}$, a contradiction. [We already saw that $\mathfrak{A}^l = a^l O = (\alpha)O$, so $a^l = (\alpha) = \alpha o$ and a is of order l.] $\qquad\square$

9.2.4a. Remark. This proves Hilbert's Theorem 94 (see 9.1.1) because $\mathfrak{A}$ must lie in o unless it has ramified prime divisors (under the assumption $\mathfrak{A} = \mathfrak{A}^s$). We have yet to prove Hilbert's Theorem 92, but we deduce one major consequence. The following result is the simplest case of Hilbert's genus theory (we otherwise cite the references):

9.2.5. Main theorem. *Let K/k be a Kummer extension that can be generated as $K = k(^l\sqrt{\mu})$ only for $\mu \notin o^*$ and for which $d_{K/k}$ has only one prime divisor. Let us further assume that the class number h of k is prime to l. Then the class number of K is prime to l.*

9.2.6. Lemma. *Under the same assumptions, the units of k are norms of units in K; that is, symbolically, $o^* = NO^*$.*

Proof of main theorem. (The proof of Lemma 9.2.6 is deferred until Lemma 9.3.5 because it follows from the mechanism of Hilbert's relative units.) We shall let g denote an integer prime to l as required. First of all, let $\mathfrak{p} = \mathfrak{P}^l$ be the one ramified prime in K/k. Then we can write $K = k(\sqrt[l]{\mu})$ for $(\mu) = \mathfrak{p}^g \mathfrak{b}^l$. Since $\mathfrak{p}^h = (\pi)$ and $\mathfrak{b}^h = (\beta)$,

$$(9.2.7a) \qquad \mu^h = \pi^g \beta^l \epsilon \qquad (\epsilon \in o^*).$$

If we raise both sides to the h' power ($hh' \equiv 1 \bmod l$) and take the lth root, we find $k(\sqrt[l]{\mu}) = k(\sqrt[l]{\mu'})$ for $g' = gh'$ and

$$(9.2.7b) \qquad \mu' = \pi^{g'} \epsilon' \qquad (\epsilon' \in o^*).$$

But $(\pi) = \mathfrak{p}^h = \mathfrak{P}^{lh}$; therefore, the ideal $\mathfrak{P}$ in K has order prime to l; that is,

$$(9.2.7c) \qquad \mathfrak{P}^{hg'} = (\sqrt[l]{\mu'}).$$

Of course, $\mathfrak{P}$ is the only prime ideal in K (and not in k) that satisfies $\mathfrak{P} = \mathfrak{P}^s$. We have to show only that the class number of K could be divisible by l only by virtue of ideal classes generated by such discriminant divisors $\mathfrak{P}$.

9.2.8. Lemma. *Assuming Lemma 9.2.6, the only ambiguous ideal classes C of K (i.e., $C = C^s$) are those that contain an ambiguous ideal $\mathfrak{A}$ (i.e., $\mathfrak{A} = \mathfrak{A}^s$, $\mathfrak{A} \neq aO$, for $a \in o$).*

Proof. Easily, if $\mathfrak{A} = \mathfrak{A}^s$, then $\mathfrak{A}$ determines $C = C^s$. Conversely, let $C = C^s$ and $\mathfrak{A} \in C$. Then $\mathfrak{A} = \mathfrak{A}^s B$ with $N(B) = (1)$. Thus, $NB = \epsilon$ (in

o^*). By Lemma 9.2.6, $\epsilon = NE$ (for E in O^*) and $NB/NE = 1$. It follows that $B/E = A^{S-1}$ for some A in K. We now have $\mathfrak{A}A$ (in C) invariant under S.

9.2.9. Lemma. *Let G be a finite abelian group of elements $\{B, C, \ldots\}$ with an automorphism of order $l(C \to C^S$ for all C in G, $S^l = 1)$. Assume the automorphism is of norm 1; that is, $NC = CC^S \cdots C^{S^{l-1}} = 1$. Then $|G| \equiv 0 \bmod l$ if and only if some element of the group satisfies $C = C^S$ for $C \neq 1$.*

Proof. Let $C = C^S \neq 1$. Then $NC = C^l = 1$, proving the "if." For the "only if," let B be of order l in G and consider the sequence

(9.2.9a) $$B^{S-1}, B^{(S-1)^2}, \ldots, B^{(S-1)^{l-1}}.$$

The element 1 must appear here. Indeed the last term is 1, since

$$(S - 1)^{l-1}(S - 1) \equiv S^l - 1 \bmod l$$
$$(S - 1)^{l-1} \equiv 1 + S + \cdots + S^{l-1} \bmod l$$

in (say) the group ring $\mathbf{Z}[S]$. Hence, we can take C as the last non-identity element of sequence (9.2.9a). $\quad\square$

Of course, we let G denote the group $\mathrm{Cl}\{K\}^h$ (the group of hth powers of the classes of k). The ideals of k go into the identity class, whereas the l-factors of $\mathrm{Cl}\{K\}$ are unchanged. This establishes the main result (Theorem 9.2.5) assuming only Lemma 9.2.6 for now. We conclude this section with a result that contrasts with Hilbert's Theorem 92.

9.2.10. Lemma. *Under the assumptions of Theorem 9.2.5, for $E \in O^*$,*

(9.2.10a) $$NE = 1 \Rightarrow E^g = \zeta^u H^{1-S}$$

for $H \in O^*$, $(g, l) = 1$, $u \in \mathbf{Z}$, and $(u, l) = 1$.

Proof. Most immediately, $NE = 1 \Rightarrow E = A^{1-S}$ for $A \in O$. Hence, if $(A) = \mathfrak{A}$, then $\mathfrak{A} = \mathfrak{A}^S$, which implies

(9.2.10b) $$\mathfrak{A} = \mathfrak{p}^x \mathfrak{b}, \qquad \mathfrak{b} \subseteq o, x \in \mathbf{Z}.$$

Therefore, by equation (9.2.7c), $\mathfrak{A}^{hg} = (\sqrt[l]{\mu'})^x \beta$ for $(\beta) = \mathfrak{b}^{hg}$ in k.

(9.2.10c) $$\mathfrak{A}^{hg} = H\mu'^{x/l}\beta.$$

This proves the required result, if we recall that $\sqrt[l]{\mu'^{1-S}}$ is an lth root of unity. $\quad\square$

As a final digression before the remaining proofs we show how ramified and unramified towers interact according to Theorems 9.1.3 and 9.2.5.

9.2.11. Illustration. If we start with a field K_1 whose class number is prime to l (and if $\zeta \in K_1$), then, inductively, the recursion

$$(9.2.11\text{a}) \qquad K_{i+1} = K_i(\sqrt[l]{\mu_i})$$

will be used to create a tower of fields with class number prime to l by ensuring that each disk K_{i+1}/K_i has one (finite) prime divisor only (as in Theorem 9.2.5).

Using the parameters of Illustration 9.1.9, we reconsider the two-class field tower ($l = 2$) for $k = \mathbf{Q}(\sqrt{-p})$, with p prime, this time showing the unit groups and ranks:

$$(9.2.11\text{b}) \qquad \begin{aligned} k &= \mathbf{Q}(\sqrt{-p}), & o^* &= \langle -1 \rangle & (r = 0) \\ k_1 &= k(\sqrt{p}), & o_1^* &= \langle i, \epsilon \rangle & (r = 1) \\ k_2 &= k_1(\sqrt{\epsilon}), & o_2^* &\underset{2}{=} \langle i, \sqrt{\epsilon}, \theta_2, \theta_2^c \rangle & (r = 3) \end{aligned}$$

$$k_3 = k_2(\sqrt{(\sqrt{\epsilon}\theta_2\theta_2^c)}),$$
$$o_3^* \underset{2}{=} \langle i, \sqrt{\epsilon}, \theta_2, \sqrt{(\sqrt{\epsilon}\theta_2\theta_2^c)}, \theta_3^{(j)}, (j = 1, \ldots, 4) \rangle \qquad (r = 7).$$

It must be explained that $\underset{2}{=}$ denotes equivalence to within odd torsion; that is, the right-hand group is a subgroup of the units of odd index. Also the symbols θ_m and indicated conjugates are a succession of units (of degrees 2^m, θ_1 is actually i) that satisfy recursive Pellian equations

$$(9.2.11\text{c}) \qquad N\theta_m = \theta_{m-1}.$$

This involves a difficult choice of conjugates, which is left to bibliographic citations.

Now we construct a ramified tower $K_1 \subset K_2 \subset \cdots$ by using subfields of k_t. Note that Gal $k_t/k = D(2^t)$ (dihedral). Thus, for each order that divides 2^{t+1}, there are three types of subfields (to within equivalence by conjugates). First there are the normal fields of the unramified tower

$$(9.2.11\text{d}) \qquad k \subset k_1 \subset k_2 \subset \cdots \subset k_t.$$

Then there are two types of subfields of each k_i: those conjugate to a real field, namely k_{iR}, and those conjugate to only complex fields, namely k_{iC}. These are always of index two; that is, $[k_i : k_{iR}] = [k_i : k_{iC}] = 2$.

Thus,

(9.2.12a)
$$k_{1R} = \mathbf{Q}(\sqrt{p}) \subset k_{2R} = \mathbf{Q}(\sqrt{\epsilon}) \subset k_{3R}$$
$$= k_{2R}(\sqrt{\sqrt{\epsilon\theta_2\theta_2^c}}) \subset \cdots$$

(9.2.12b)
$$k_{1C} = \mathbf{Q}(i) \subset k_{2C} = k_{1C}(\sqrt{\pi}) \subset \cdots \qquad (\pi\pi' = p).$$

Now we have the tower (9.2.12b) of *ramified* quadratic extensions [within the two-class field tower (9.2.11d)]. The adjunctions satisfy the conditions of Lemma 9.2.10 so they have odd class number, and the units satisfy Lemma 9.2.6.

9.3. Hilbert's relative units

We conclude by outlining Hilbert's theory of relative units, which is used to finally prove Hilbert's Theorem 92 (Theorem 9.2.3) and Lemma 9.2.6.

9.3.1a. Definition. *Let K/k be a cyclic extension of prime degree 1, not necessarily Kummerian, but unramified at ∞. For k, the torsion-free rank of o^* is $r = r_1 + r_2 - 1$, whereas for K the rank of O^* is $R = l(r + 1) - 1$. We define a new group of units $[o^*]$, a subgroup of O^* consisting of those whose lth power lies in o^*. Thus, $o^* = [o^*]$, except for the case where $K = k(\sqrt[l]{\epsilon})$ for ϵ in o^*, in which case $|[o^*]/o^*| = l$. (As usual, we use each of these set symbols to denote an element of the set when the context is clear.) We call the units of K*

(9.3.1b)
$$H_1, \ldots, H_{r+1}$$

"relative units" for K/k when they are "minimal" for representing the subgroup E^{1-S} (for E in O^*) in the following sense: Let $F_i(S) \in Z[S]$ (where S, as usual, is a generator of Gal K/k). Then by definition,

(9.3.1c)
$$H_1^{F_1(S)} \cdots H_{r+1}^{F_{r+1}(S)}[o^*] = E^{1-S}$$

if and only if each $F_i(\zeta)$ is divisible by $1 - \zeta$. Here ζ is a primitive lth root of unity (not necessarily in k), so equivalently, the condition is that $F_i(1)$ be divisible by l.

9.3.2. Theorem. *Hilbert's relative units exist and, indeed, their conjugates under Gal K/k together with an independent set for k form an independent set of units of K.*

Proof. This is seen directly in the construction. We start with r torsion-free fundamental units of o^* and augment them by $R - r = (r + 1)(l - 1)$ units of K:

$$(9.3.2a) \qquad E_1, \ldots, E_1^{S^{l-2}}, \ldots, E_{r+1}, \ldots, E_{r+1}^{S^{l-2}}$$

chosen as follows: We choose E_1 in $O^*\backslash[o^*]$. Then all $l - 1$ (indicated) conjugates of E_1 are independent over $[o^*]$. This is so because if $E_1^{F(S)}$ lies in $[o^*]$, it also lies in o^* for some $F(S) \in \mathbf{Z}[S]$. To complete the independence proof, note first that $F(S)$ can be taken to be of degree less than or equal to $l - 2$ and relatively prime to $\Phi(S) = 1 + S + \cdots + S^{l-1}$. Therefore, we can write $F(S)X(S) + \Phi(S)Y(S) = f$ for $X(S)$ and $Y(S)$ in $\mathbf{Z}[S]$ and f in $\mathbf{Z}$. Thus, $E_1^{f(S)}$ and consequently E_1^f lie in o^*, contradicting the choice of E_1. If O_1^* is the new unit group generated by $[o^*]$ and E_1 (with conjugates), then we choose E_2 in $O^*\backslash O^*_1$ to form O_2^*, the group generated using conjugates, and proceed until we exhaust the rank R by adjoining the set (9.3.2a) and forming O^*_{r+1}, a group of units with rank R. Unfortunately, we cannot find fundamental units while insisting that each unit in the basis be included with all its conjugates. We do consider the $R - r$ units of (9.3.2a) mod $[o^*]$ as a Galois module over $E_1, \ldots, E_{r+1}$ under the ring of $\mathbf{Z}[S]/\Phi(S)$. We reduce those units to a minimal set with regard to divisors of $(1 - S)$ in the same way that the factors of p^2 (for p prime) are removed from the discriminant by modifying the integer basis of a submodule of O^* (see Section 4.1). Thus for each t with $1 \le t \le r + 1$, we choose a unit H_t so that for maximal $e \ge 0$,

$$(9.3.2b) \qquad E_t^{F_t(S)} \cdots E_{r+1}^{F_{r+1}(S)}[o^*] = H_t^{(1-S)^e}$$

where not all $F_j(1)$ values are divisible by l. $\qquad\qquad \square$

We have a new set of units generated from Hilbert relative units:

$$(9.3.2c) \quad O_H^* = \langle H_t^{S^j}, [o^*] \rangle, \qquad 1 \le t \le r + 1, \qquad 0 \le j \le l - 2.$$

9.3.3. Theorem. *The index $j = |O^*/O_H^*|$ is prime to l; that is, for an arbitrary E in O^* and this same j, there exists $F_i(S)$ in $\mathbf{Z}[S]$ such that*

$$(9.3.3a) \qquad E^j = H_1^{F_1(S)} \cdots H_{r+1}^{F_{r+1}(S)}[o^*].$$

Thus, for every E in $O^*/[o^*]$, E^j corresponds uniquely to a vector of elements of $\mathbf{Z}[S]/\Phi(S)$, namely $[F_1(S), \ldots, F_{r+1}(S)]$.

Proof. Note that if $l \mid j$, then the factor l can be removed. This is true because $l = (1 - \zeta) \cdots (1 - \zeta^{l-1})$, so, for example, E^l is $E^{(1-S)\cdots(1-S^{l-1})}$ mod $[o^*]$. The exponent l is then removable in j by the definition of relative units. □

9.3.3b. Corollary. *If $\epsilon = NH$ for some H in O^*, then with $\eta_i = NH_i$,*

$$\epsilon = \eta_1^{F_1(1)} \cdots \eta_{r+1}^{F_{r+1}(1)}[o^*]^l.$$

This result follows from taking the norm of equation (9.3.3a). Of course, $1 - S \mid F_i(1) - F_i(S)$, so ϵ^j is clearly of the form shown. The euclidean algorithm reduces the exponent from j to 1, since $(l, j) = 1$.

9.3.4 *Proof of Hilbert's Theorem 92 (Theorem 9.2.3).* The basic method is most easily illustrated when ζ is not in k. Then if we define

(9.3.4a) $$\eta_i = NH_i \qquad (1 \leq i \leq r + 1)$$

some identity of the type

(9.3.4b) $$\eta_1^{a_1} \cdots \eta_{r+1}^{a_{r+1}} = 1$$

must follow because the units have rank r. If all a_i are divisible by l, we could take an lth root, knowing that $1^{1/l} = 1$ only. Thus if (say) $(a_{r+1}, l) = 1$, we can set

(9.3.4c) $$H = H_1^{a_1} \cdots H_{r+1}^{a_{r+1}}.$$

Then from equation (9.3.4b), $NH = 1$, and from the definition of relative units, $H \neq E^{1-S}$.

The proof becomes more involved if ζ lies in k. We note that $N\zeta = 1$, but $\zeta = E^{1-S}$ only when $E = \epsilon^{1/l}$. This is now the hard case. We set $\eta_{r+2} = E^l \; (= \epsilon = NE)$ and let ζ' be the maximal primitive l^h root of unity (ζ' lies in k but its lth root does not). We find a new relation of the following type, but with not all a_t divisible by l:

(9.3.4d) $$\eta_1^{a_1} \cdots \eta_{r+2}^{a_{r+2}} = \zeta'^{al}.$$

(Since the unit rank is r, there are two relations that can be manipulated to make the exponent of ζ' divisible by l.) Now define

(9.3.4e) $$H = H_1^{a_1} \cdots H_{r+1}^{a_{r+1}} E^{a_{r+2}} \zeta'^{-a} = H_1^{a_1} \cdots H_{r+1}^{a_{r+1}}[o^*].$$

Clearly, $NH = 1$. We must show that it is not possible for all a_t to be divisible by l for $1 \leq t \leq r + 1$ [so we could proceed as in equation

(9.3.4c)]. Thus, if all $a_1, \ldots, a_{r+1}$ were divisible by l (so a_{r+2} is not), we could take an lth root of equation (9.3.4d):

$$(9.3.4f) \qquad \eta_1^{a_1/l} \cdots \eta_{r+1}^{a_{r+1}/l} E^{a_{r+2}} = \zeta^{\prime a} \zeta^b.$$

This is a contradiction because E does not lie in k.   □

We now complete the proof of Lemma 9.2.6 in the following equivalent form:

9.3.5. Lemma. *Assume the hypotheses of the main theorem 9.2.5. Then under the norm mapping, the equivalence classes O^*/O^{*l} project surjectively upon the equivalent classes o^*/o^{*l}. [Indeed, the same is true if we restrict O^* to O_H^*, defined in equation (9.3.2c).]*

Proof. From the definition of Hilbert units,

$$O_H^*/O_H^{*l} = H_1^{F_1(S)} \cdots H_{r+1}^{F_{r+1}(S)}\{\epsilon\}$$

where $F_i(S)$ are polynomials of degree $l - 1$ in $\mathbf{Z}[S]$ mod l and $\{\epsilon\}$ is an equivalence class of o^*/o^{*l}. Thus, there are $(l - 1)(r + 1)$ replicas of $\mathbf{Z}/l\mathbf{Z}$ in the norm mapping shown. The image o^*/o^{*l} likewise consists of $(r + 1)$ replicas by the independence of fundamental units (with torsion). We ask which elements of O_H^*/O_H^{*l} are in the kernel, or (by Corollary 9.3.3b) when is $NE = \epsilon^l$ for units of K and k? Clearly, by Lemma 9.2.10,

$$(E/\epsilon)^g = \zeta H^{1-S}, \qquad H \in O^*, \qquad (g, l) = 1.$$

Thus, by the definition of Hilbert units, E belongs to the kernel precisely when $(1 - S)$ divides each polynomial $F_i(S)$. This fact is determined by a single coefficient modulo l. Thus, the cokernel consists of $(r + 1)$ replicas of $\mathbf{Z}/l\mathbf{Z}$ [and the kernel of $(r + 1)(l - 2)$ replicas.] The dimensions match and the surjectivity is proved.   □

9.3.6. Remark. The special form of Lemma 9.3.5 is interesting from an algebraic point of view, but we need only the consequence

$$o^* = N(O_H^*) = N(O^*).$$

This follows immediately from the surjectivity because for any $\eta \in o^*$, there exist $\eta_0 \in o^*$ and E, E_0 in O^* so that $\eta/\eta_0^l = N(E/E_0^l)$; hence, $\eta = N(E\eta_0/E^l)$. Note that even though O_H^* may be of finite index g [with $(g, l) = 1$] in O^*, the norm mapping produces the same image o^*.

9.4. The governing field

We conclude this chapter on the fields defined by radicals with some related illustrations that may, in a sense, extend the objectives of class field theory. We define a fixed field that governs the class number and structure of an arbitrary field according to the splitting of primes in the discriminant of this field. This extends the concept of a class field, where the factorization of an arbitrary prime is governed by its ideal class.

9.4.1. Definition. *The normal field Ω is said to be a "governing field" for class structure $\mathbf{Z}/H_1\mathbf{Z} \times \cdots \times \mathbf{Z}/H_t\mathbf{Z}$ and multiplier M precisely when (with finite exceptions) the splitting of a prime q in $\Omega/\mathbf{Q}$ is sufficient for the field $\mathbf{Q}(\sqrt{Mq})$ to have a (strict) ideal group with $H = \mathbf{Z}/H_1\mathbf{Z} \times \cdots \times \mathbf{Z}/H_t\mathbf{Z}$ as a subgroup. The minimal such field (i.e., the intersection of all such fields) is called "the" governing field and is denoted by $\Omega_M[H_1, \ldots, H_t]$ (or merely $\Omega[H_1, \ldots, H_t]$ if M is understood).*

The concept was introduced by Lagarias and Morton, who used a more flexible definition, in which q does not always split but has prescribed Artin characters. There are ways to devise a more general definition, but this task might prove premature in view of the current paucity of illustrations. It is not even clear when such an Ω exists, and indeed a further search for examples might well lead to an unpredictable revision of the definition. The strongest evidence for Ω is that the class groups under discussion occur with an observable Dirichlet density that is easily attributable to a Cheboratev density and hence (we might hope) to an Artin symbol.

9.4.2. Illustrations from genus theory. If $\pm M$ is not a square then we note $\Omega[2] = \mathbf{Q}$. [All fields $\mathbf{Q}(\sqrt{Mq})$, with $(M, q) = 1$, have two different primes in the discriminant and hence an even class number (at least in the strict sense).] Likewise, $\Omega[4] \subseteq k_g(i)$, where k_g is the genus field for $\mathbf{Q}(\sqrt{M})$. This last result follows from the condition for the prime divisor of q ($= q_1^2$) to be in a squared class in $\mathbf{Q}(\sqrt{Mq})$. Further refinements and illustrations of $\Omega[2, \ldots, 4, \ldots]$ can be made accordingly.

9.4.3. Special illustrations for $M = -4$. Here we build on a succession of classical results:

(9.4.3a) $\Omega[2] = \mathbf{Q}(i)$

since $\mathbf{Q}(\sqrt{-q})$ has an even class number precisely when q splits in $\mathbf{Q}(i)/\mathbf{Q}$, or $q \equiv 1 \bmod 4$. Furthermore,

(9.4.3b) $\Omega[4] = \mathbf{Q}(i, \sqrt{2}) = \mathbf{Q}(\zeta)$ ($\zeta = \exp 2\pi i/8$)

that is, $\mathbf{Q}(\sqrt{-q})$ has a class of order 4 precisely when $q \equiv 1 \bmod 8$. A more recent result of Barrucand and Cohn is

(9.4.3c) $\Omega[8] = \mathbf{Q}(\zeta, \sqrt{1 + \sqrt{2}})$

(with the equivalent splitting condition that $q = a^2 + 32b^2$). We might expect a similar result on $\Omega[16]$, but observe the following:

9.4.4. Lemma (Cohn–Lagarias). *Either $\Omega[16]$ does not exist or else it cannot be a normal field of degree 16 with even ramifications only.*

Proof. Assume $\Omega[16]$ to exist contrary to hypothesis. Then $\Omega[16] \supset \Omega[4]$ so Gal $\Omega[16]/\Omega[4]$ is either $\mathbf{Z}/2\mathbf{Z} \times \mathbf{Z}/2\mathbf{Z}$ or $\mathbf{Z}/4\mathbf{Z}$. In these cases, we need to recall facts about the conductor $\mathfrak{f}$, the discriminant d, and different δ for quadratic and biquadratic extensions:

1. If $[K : k] = 2$; then by Illustration 7.4.14 $d_{K/k} = \mathfrak{f}(K/k) = \delta_{K/k}^2$.
2. If $[K : k] = 2$ and K/k has only even ramifications whereas 2 is totally ramified in $k/\mathbf{Q}$, then by Lemma 6.2.16a $\delta_{K/k} \mid 2 \cdot 2_{11}$, where $2_{11}^2 = 2_1$, the divisor of 2 in k. Thus, $d_{K/k} \mid 4 \cdot 2_1$. [In essence, $k(\sqrt{2_1})$ is the worse type of 2-ramification.]
3. If $K \supset k_1 \supset k$ is a tower of quadratic extensions, then by Theorem 6.1.11

$$\delta_{K/k_i} \cdot \delta_{k_i/k} = \delta_{K/k}.$$

4. If Gal $K/k = \mathbf{Z}/2\mathbf{Z} \times \mathbf{Z}/2\mathbf{Z}$ with k_1, k_2, and k_3 the intermediate fields between K and k, then by Corollary 7.4.9

$$\mathfrak{f}(K/k) = \mathrm{lcm}\ [\mathfrak{f}(k_t/k)], \qquad t = 1, 2, 3.$$

5. If Gal $K/k = \mathbf{Z}/4\mathbf{Z}$ with $K \supset k_1 \supset k$, then by Exercise 7.4

$$d_{K/k} = \mathfrak{f}(K/k)^2 \mathfrak{f}(k_1/k)$$
$$\mathfrak{f}(K/k)^2 = d_{K/k_1}^2\, d_{k_1/k}.$$

In the present context of (9.4.3a, b, c), for $\Omega[4]$,

(9.4.4a) $$(1 + \zeta)^4 = (2)$$

and $\Omega[8]/\Omega[4]$ is generated by an element of minimal relative discriminant, $(1 + \sqrt{1 + \sqrt{2}})/(1 + \zeta)$, so

(9.4.4b) $$\delta_{\Omega[8]/\Omega[4]} = 2/(1 + \zeta) = (1 + \zeta)^3.$$

In $\Omega[8]$, we have the total ramification of 2; that is,

(9.4.4c) $$(\Lambda)^8 = (2), \qquad \Lambda = (1 + i)^{-1} - \zeta/(1 + \zeta).$$

Therefore $d_{\Omega[16]/\Omega[8]}$ divides 4Λ by condition 2.
Also, from conditions 2 and 4, we find

(9.4.5a) $$\mathfrak{f}(\Omega[16]/\Omega[4]) \mid 4(1 + \zeta) = (1 + \zeta)^9.$$

Meanwhile, proceeding from condition 5, we would find

(9.4.5b) $$\mathfrak{f}(\Omega[16]/\Omega[4])^2 \mid (4\Lambda)^2(1 + \zeta)^6 = (1 + \zeta)^{23}.$$

By either condition 4 or 5, since $\mathfrak{f}(\Omega[16]/\Omega[4])$ is an exact power of $(1 + \zeta)$ [compare equation (9.4.4a)],

(9.4.6) $$\mathfrak{f}(\Omega[16]/\Omega[4]) \mid (1 + \zeta)^{11}.$$

We can now complete the proof by using counterexamples of primes factored in $\Omega[4]$:

$$-11 + 12\zeta^2 + 16\zeta^3 = \pi_{593} \mid 593$$
$$-7 + 4\zeta + 4\zeta^3 = \pi_{8273} \mid 8273$$
$$\pi_{593} - \pi_{8273} \equiv 4(1 + \zeta)^3 \bmod 8 \equiv 0 \bmod \mathfrak{f}(\Omega[16]/\Omega[4]).$$

Therefore, if one of the primes $q = 593$ or 8273 splits from $\mathbf{Q}$ to $\Omega[16]$, then so must the other. Now it can be seen that the governing property is contradicted as follows:

$$|\operatorname{Cl}\{\mathbf{Q}(\sqrt{-593})\}| = 24, \qquad |\operatorname{Cl}\{\mathbf{Q}(\sqrt{-8273})\}| = 64$$

(only one class number is divisible by 16). $\qquad\square$

9.4.7. Remark. A direct but more laborious method is to list all possible quadratic extensions of $\Omega[8]$ that ramify only at 2 and then to find primes like the preceding pair that split alike for each extension while the fields display contradictory class structures. The details are given in the references, but we can give the following stronger result achieved by this direct method (and stated here without proof).

9.4.8. Lemma. *Either $\Omega[16]$ does not exist for $M = -4$ or else it can not be in a normal field of degree 32 of even ramifications that contains the sixteenth roots of unity. (Such a field would also have to contain $\Omega[8]$ as well as $2^{1/4}$, $\sqrt{1+i}$, $\sqrt{2+\sqrt{2}}$, and many other "reasonable" quantities.)*

In summary, the governing field may be seen as inchaote today as the class field must have seemed over a century ago. Classically, the problem begins with the representation (in $\mathbf{Z}$) of $m = F_d(x, y)$ for the principal quadratic form of given discriminant d. The first step (and indeed the crucial step) was to take $m = p$, a prime, and find the field $K (= \text{RCF}\{d\})$ over $k = \mathbf{Q}(\sqrt{d})$ such that (wfe)

$$(9.4.8a) \qquad\qquad p \in \text{Spl}(K) \Leftrightarrow p = F_d(x, y).$$

This process ultimately involves the (ring) class structure of k as seen in $G = \text{Gal } K/k$. For instance, m can be represented by $F_d(x, y)$ even if some of its prime factors do not satisfy equation (9.4.8a) as long as the product of the ideal factors is principal.

A new question is now posed with d as a variable and some group structure $H = \mathbf{Z}/\mathbf{Z}H_1 \times \cdots \times \mathbf{Z}/\mathbf{Z}H_t$ prescribed: For which d does $G = \text{Gal RCF}\{d\}/\mathbf{Q}(\sqrt{d}) \supseteq H$? As before, we reduce to a single factor of interest in d, $d = Mq$ for q prime, and we ask for a field such that

$$(9.4.8b) \qquad\qquad q \in \text{Spl}(\Omega) \Leftrightarrow G \supseteq H.$$

Yet this question is not so far from questions about class fields. For instance, if

$$(9.4.8c) \quad d = -4q, \qquad M = -4, \qquad F_d(x, y) = x^2 + qy^2$$

with $H = \mathbf{Z}/2^s\mathbf{Z}$, we have the result

$$(9.4.8d) \qquad \begin{aligned} q \in \text{Spl}(\Omega) \Leftrightarrow \{pz &= F_d(x, y) \Rightarrow \\ p^{2^s} &= F_d(x', y'), \qquad (x, y) = (x', y') = 1\} \end{aligned}$$

for $\Omega = \Omega_M[2^s]$. Indeed, this Ω exists for $s = 0, 1, 2$. Therefore the governing field continues to pursue objectives analogous to those of a class field but on a slightly different level [as we might conclude by comparing the relations (9.4.8a) and (9.4.8d)].

Exercises

9.1. Show that for $l = 2$, in the definition of Hilbert relative units (9.3.1c) can be replaced by

$$H_1^{e_1} \cdots H_{r+1}^{e_{r+1}}[o^*] = E^2 \Rightarrow \text{(each) } e_i \equiv 0 \bmod 2.$$

Verify the equivalent definition:

$$H_1^{e_1} \cdots H_{r+1}^{e_{r+1}}[o^*] = E^u \qquad (u \text{ odd})$$

where E is a general unit of K and e_i and u are determined by E. Note that in the first definition we can replace $[o^*]$ by o^*, but not in the second definition.

9.2. If l is odd, show that the replacement of $[o^*]$ by o^* (cited in Exercise 9.1) would not work in definition (9.3.1c). (*Hint:* $l \neq 1 - \zeta$.)

9.3. If l is an odd prime and $K/\mathbf{Q}$ is cyclic of degree l, then there exists a so-called normal unit E in K such that the conjugates of E generate a subgroup of units in O^* of index prime to l.

9.4. Verify these cases:

a. $k = \mathbf{Q}$, $K = \mathbf{Q}(\sqrt{d})$, $d > 0$ (only), $H_1 = \eta$, $(\eta \in O^*, \sqrt{\pm\eta} \notin O^*)$

b. $k = \mathbf{Q}(\sqrt{3})$, $K = k(\sqrt{2})$, $H_1 = 1 + \sqrt{2}$, $H_2 = \sqrt{2} + \sqrt{3}$, $[o^*] = \langle -1, (1 + \sqrt{3})/\sqrt{2} \rangle$

c. $k = \mathbf{Q}(i)$, $K = k(\sqrt{2})$, $H_1 = 1 + \sqrt{2}$, $[o^*] = \langle (1 + i)/\sqrt{2} \rangle$

d. $k = \mathbf{Q}(\sqrt{2})$, $K = k(\sqrt{2 + \sqrt{2}})$, $H_1 = 1 + \sqrt{2 + \sqrt{2}}$, $H_2 = 1 + \sqrt{2 - \sqrt{2}}$, $[o^*] = \langle -1, 1 + \sqrt{2} \rangle$

e. $K = \mathbf{Q}$, $K = \mathbf{Q}(\zeta + 1/\zeta)$, $H_1 = \zeta + 1/\zeta$, $[o^*] = \langle -1 \rangle$ ($\zeta = \exp 2\pi i/7$)

9.5. Show that the system of definitions of relative units breaks down with ramifications at ∞ for K/k. For $k = \mathbf{Q}(\sqrt{d})$, $d > 0$, and $K = k(i)$, show there are cases where Theorem 9.2.5 and Lemma 9.2.6 fail. [*Hint:* $|\mathrm{Cl}\{\mathbf{Q}(\sqrt{-17}\}| = 4$ and $|\mathrm{Cl}\{\mathbf{Q}(\sqrt{-17}, \sqrt{17})\}| = 2$.]

9.6. Apply Theorem 9.2.5 and Lemma 9.2.6 to $\mathbf{Q}(\sqrt{d})/\mathbf{Q}$ and the Pellian equation.

Bibliographic note

For background theory on towers, see Furtwängler (1916), Hilbert (1897), Kisilevsky (1970, 1976), and Taussky (1937, 1971). For group-theoretic results, see Hall (1959) and for theorems on units, see Hilbert (1897, 1899). Some applications to quadratic towers are detailed in Cohn and Cooke (1976) and Cohn (1979c, 1981a). The concept of governing field appears in Morton (1982) and in Cohn and Lagarias (1983).

10

The modular function field

In Chapter 5 the modular function of Klein $J(\tau)$ [or Weber $j(\tau)$] was introduced as an invariant of both the cubic curve and the period lattice of the elliptic integral to which the curve leads. Specifically, we take the curve

$$(10.0.1) \quad y^2 = 4x^3 - g_2 x - g_3 = 4(x - r_1)(x - r_2)(x - r_3)$$

factored to show roots, which are assumed unequal. Thus,

$$(10.0.2) \quad 16[(r_1 - r_2)(r_2 - r_3)(r_3 - r_1)]^2 = g_2^3 - 27g_3^2 \neq 0.$$

Then the elliptic integral (of the first kind)

$$(10.0.3) \qquad\qquad u = \int dx/y$$

over closed paths on the complex x-sphere generates a doubly periodic structure. The periods form a lattice

$$(10.0.4) \quad \mathcal{L} = [2\omega_1, 2\omega_2] = \{\Omega\}, \qquad \Omega = 2\omega_1 n_1 + 2\omega_2 n_2$$

where $n_1, n_2 \in \mathbf{Z}$ and $\operatorname{Im} \tau > 0$ for

$$(10.0.5) \qquad\qquad \tau = \omega_1/\omega_2.$$

The coefficients g_2 and g_3 are determined from $\mathcal{L}$ by summations:

$$(10.0.6) \qquad g_2 = 60 \, \Sigma' \, \Sigma^{-4}, \qquad g_3 = 140 \, \Sigma' \, \Omega^{-6}.$$

Then the modular functions are defined by

$$(10.0.7) \qquad j(\tau) = [1728J(\tau) =]1728g_2^3/(g_2^3 - 27g_3^2).$$

It turns out to be a relatively simple matter to see that $j(\tau)$ has *rational* coefficients in its q-expansion, that is, its expansion in the variable

153

$q = \exp 2\pi i\tau$. It is, however, imperative to show that these coefficients are *integral*. [This integrality property makes $j(\tau)$ preferable to $J(\tau)$ for number-theoretic purposes.] To demonstrate the integrality of these coefficients, we shall invoke the use of algebraic function theory on Riemann surfaces, including the use of differentials (or modular forms). The most important one is the "discriminant":

$$(10.0.8) \qquad \Delta(\tau) = (g_2^3 - 27g_3^2)(\omega_2/\pi)^{12}.$$

10.1. Summation formulas

We are dealing with Eisenstein series G_{2k} (i.e., $g_2 = 60G_4$ and $g_3 = 140G_6$). In general,

$$(10.1.1) \qquad G_{2k}(\omega_1, \omega_2) = \Sigma' \, (2n_1\omega_1 + 2n_2\omega_2)^{-2k}, \qquad k \geq 2.$$

(The sum is over integer pairs n_1, n_2 not both zero.) It is convenient to express the function $G_{2k}(\tau)$ for $\tau = \omega_1/\omega_2$ (Im $\tau > 0$) as follows:

$$(10.1.2) \qquad G_{2k}(\omega_1, \omega_2)(2\omega_2)^{2k} = G_{2k}(\tau) = \Sigma' \, (n\tau + m)^{-2k}$$

over the integral pairs m, n not both zero. With $k \geq 2$, we have a uniform convergence that easily establishes the two identities:

$$(10.1.3) \quad G_{2k}(-1/\tau) = \tau^{2k}G_{2k}(\tau), \qquad G_{2k}(\tau + 1) = G_{2k}(\tau), \qquad k \geq 2$$

by rearranging summands. [In the first case $(m, n) \rightarrow (-n, m)$ and in the second $(m, n) \rightarrow (m + n, n)$.]

If we introduce the new variable

$$(10.1.4) \qquad\qquad q = \exp 2\pi i\tau$$

it is clear that (by the periodicity in τ) $G_{2k}(\tau)$ is single-valued in q and therefore expressible by a Laurent series in q. This series can be found rather easily by first writing

$$(10.1.5) \qquad G_{2k}(\tau) = 2 \sum_{m=1}^{\infty} m^{-2k} + 2 \sum_{n=1}^{\infty} \sum_{m=-\infty}^{\infty} (n\tau + m)^{-2k}.$$

We start with the formula of Euler:

$$\sum_{m=-\infty}^{\infty} (\tau + m)^{-2} = \pi^2 \csc^2 \pi\tau$$

$$= (2\pi i)^2/(\exp \pi i\tau - \exp -\pi i\tau)^2 = (2\pi i)^2 q/(1 - q)^2$$

which expands into

$$\sum_{m=-\infty}^{\infty} (\tau + m)^{-2} = (2\pi i)^2 \sum_{h=1}^{\infty} hq^h.$$

By repeatedly differentiating with respect to τ ($2k - 2$ times), we find

(10.1.6) $\displaystyle (-1)^k(2k - 1)! \sum_{m=-\infty}^{\infty} (\tau + m)^{-2k} = (2\pi i)^{2k} \sum_{h=1}^{\infty} h^{2k-1}q^h.$

To apply this formula to equation (10.1.5), we substitute $n\tau$ for τ (i.e., q^n for q). We find

(10.1.7)

$$G_{2k}(\tau) = 2 \sum_{m=1}^{\infty} m^{-2k} + [2(2\pi)^{2k}(-1)^k/(2k - 1)!] \sum_{n=1}^{\infty} \sum_{h=1}^{\infty} h^{2k-1}q^{nh}.$$

We introduce the usual notations, $\zeta(2k)$, and the divisor sums

(10.1.8) $$\sigma_r(N) = \sum_{h|N} h^r.$$

Thus, $\sigma_r(1) = 1$, $\sigma_r(2) = 1 + 2^r$, $\sigma_r(3) = 1 + 3^r$, $\sigma_r(4) = 1 + 2^r + 4^r$, and so on. We now have a new "normalized" function $E_{2k}(\tau)$ given by

(10.1.9a) $G_{2k}(\tau) = 2\zeta(2k)E_{2k}(\tau)$ $(k \geq 2)$

(10.1.9b) $\displaystyle E_{2k}(\tau) = 1 + c_{2k} \sum_{N=1}^{\infty} \sigma_{2k-1}(N)q^N$

(10.1.9c) $c_{2k} = (2\pi)^{2k}(-1)^k/[\zeta(2k)(2k - 1)!].$

We must next evaluate $\zeta(2k)$. This is done by using Bernoulli numbers B_n and Bernoulli polynomials $B_n(x)$. They are defined inductively as follows:

(10.1.10)
$$B_0(x) = 1$$
$$B_n'(x) = nB_{n-1}(x)$$
$$\int_0^1 B_n(x)\, dx = 0, \qquad (n > 0).$$

Furthermore, $B_n(0) = B_n$. Thus,

(10.1.11)
$$
\begin{aligned}
B_0(x) &= 1, & B_0 &= 1 \\
B_1(x) &= x - \tfrac{1}{2}, & B_1 &= -\tfrac{1}{2} \\
B_2(x) &= x^2 - x + \tfrac{1}{6}, & B_2 &= \tfrac{1}{6} \\
B_3(x) &= x^3 - 3x^2/2 + x/2, & B_3 &= 0 \\
B_4(x) &= x^4 - 2x^3 + x^2 - \tfrac{1}{30}, & B_4 &= -\tfrac{1}{30}.
\end{aligned}
$$

Likewise $B_6 = \frac{1}{42}$, $B_8 = -\frac{1}{30}$, $B_{10} = \frac{5}{66}$, $B_{12} = -\frac{691}{2730}$, and so on, whereas $B_{2m+1} = 0$ for $m > 0$. Various properties of the Bernoulli polynomials and numbers can be deduced from

$$(10.1.12) \qquad u/(\exp u - 1) = \sum_{n=0}^{\infty} u^n B_n/n!$$

(see Exercise 10.1).

What concerns us now is a pair of Fourier expansions that are valid for $0 < x < 1$, namely,

$$(10.1.13a) \quad B_{2m+1}(x) = (-1)^{m-1} 2(2m + 1)!(2\pi)^{-2m-1} \sum_{n=1}^{\infty} \sin 2\pi nx/n^{2m+1}$$

$$(10.1.13b) \quad B_{2m}(x) = (-1)^{m-1} 2(2m)!(2\pi)^{-2m} \sum_{n=1}^{\infty} \cos 2\pi nx/n^{2m}.$$

If $m \geq 1$, then both series are absolutely convergent for all x, and the series represents a periodic continuation of $B_n(x)$ for all x, with period unity. But we might just as easily start with $B_1(x) = x - \frac{1}{2}$. We can expand it as a Fourier series, although the series is not absolutely convergent. We use the following theorem.

10.1.14. Theorem of Fourier–Poisson–Dirichlet. *Let $f(t)$ be a (complex) function of a real t with continuous derivative for $0 \leq t \leq 1$ and let it be extended periodically as $f_*(t)$, where*

$$(10.1.14a) \qquad \begin{aligned} f_*(0) &= f(1) = [f(0) + f(1)]/2 \\ f_*(t) &= f(t), \qquad 0 < t < 1 \\ f_*(t + n) &= f_*(t), \qquad n \in \mathbf{Z}. \end{aligned}$$

Then

$$(10.1.14b) \qquad f_*(t) = \sum_{M=-\infty}^{\infty} A_M \exp 2\pi i M t$$

where

$$A_M = \int_0^1 f_*(t)(\exp -2\pi i M t) \, dt.$$

Here the series converges for all t and indeed uniformly in any closed interval that does not contain an integer. If $f_*(0) = f_*(1) = f(0) = f(1)$, *then it converges uniformly for all t. In particular, at t = 0, the average of f(t) and f(1 − t) yields*

(10.1.14c) $[f(0) + f(1)]/2 = \sum_{M=-\infty}^{\infty} \int_0^1 f(t)(\cos 2\pi Mt)\, dt.$

We see that equation (10.1.14b) leads to the required expansion of $B_1(x)$, and all of equations (10.1.13a, b) follow. Hence for $k \geq 1$ with $x = 0$,

(10.1.15) $\zeta(2k) = B_{2k}(-1)^{k-1}(2\pi)^{2k}/2(2k)!$

so $\zeta(2) = \pi^2/6$, $\zeta(4) = \pi^4/90$, $\zeta(6) = \pi^6/945$, and so on. For now we have the rational expansions (q-series):

(10.1.15a) $E_{2k}(\tau) = 1 + c_{2k} \sum_{n=1}^{\infty} \sigma_{2k-1}(n)q^n, \qquad k \geq 2$

(10.1.15b) $c_{2k} = -4k/B_{2k}$

so $c_4 = 240$, $c_6 = -504$, $c_8 = 480$, $c_{10} = -264$, $c_{12} = 65{,}520/691$, and so on. (Note that the c_{2k} need not be integral.) At any rate, $E_4(\tau)$ and $E_6(\tau)$ have *integral* q-expansions. Also we have a formula for $j(\tau)$ that displays at least rational q-expansions:

(10.1.16) $j(\tau) = 1728E_4^3/(E_4^3 - E_6^2).$

From equation (10.1.15a) it is easy to verify a few starting terms for

(10.1.17) $E_4^3 - E_6^2 = 1728(q - 24q^2 + 252q^3 + \cdots).$

If we can show that all coefficients are divisible by 1728, then $j(\tau)$ has an *integral* q-expansion:

(10.1.18)
$$\begin{aligned}
j(\tau) = {}& 1/q + 744 + 196{,}884q + 21{,}493{,}760q^2 \\
& + 864{,}299{,}970q^3 + 20{,}245{,}856{,}256q^4 \\
& + 333{,}202{,}640{,}600q^5 + 4{,}252{,}023{,}300{,}096q^6 + \cdots.
\end{aligned}$$

(This formula is sufficiently precise to verify most numerical evaluations satisfactorily, with $\text{Im } \tau \geq \sqrt{3}/2$ and $|q| \leq 0.004333 \ldots$.) The integrality property is a major motivation of this chapter.

A necessary first step is to redefine $G_{2k}(\tau)$ for $k = 1$, where absolute convergence fails in equation (10.1.2). We use the special arrangement in equation (10.1.5) as a definition. Thus, from equation (10.1.7),

(10.1.19a) $G_2(\tau) = 2\zeta(2) + 2 \sum_{n=1}^{\infty} \sum_{m=-\infty}^{\infty} (n\tau + m)^{-2}$

$$(10.1.19b) \qquad G_2(\tau) = \pi^2/3 - 8\pi^2 \sum_{n=1}^{\infty} \sum_{h=1}^{\infty} h q^{nh}$$

$$= \pi^2/3 - 8\pi^2 \sum_{h=1}^{\infty} \sum_{n=1}^{\infty} h q^{nh}$$

$$(10.1.19c) \qquad G_2(\tau) = \pi^2/3 - 8\pi^2 \sum_{h=1}^{\infty} h q^h/(1 - q^h).$$

[There is no difficulty in interchanging sums over h and n as there was in interchanging over n and m in equation (10.1.19a).] We are assured that under definition (10.1.19a), $G_2(\tau + 1) = G_2(\tau)$. We must compare, however, $G_2(\tau)$ and $G_2(-1/\tau)/\tau^2$, which follow:

$$(10.1.20a) \qquad G_2(\tau) = 2 \sum_{m=1}^{\infty} m^{-2} + 2 \sum_{n=1}^{\infty} (n\tau)^{-2}$$

$$+ 2 \sum_{n=1}^{\infty} \sum_{m=1}^{\infty} [(n\tau + m)^{-2} + (n\tau - m)^{-2}]$$

$$(10.1.20b) \qquad G_2(-1/\tau)/\tau^2 = 2 \sum_{n=1}^{\infty} n^{-2} + 2 \sum_{m=1}^{\infty} (m\tau)^{-2}$$

$$+ 2 \sum_{n=1}^{\infty} \sum_{m=1}^{\infty} [(n - m\tau)^{-2} + (n + m\tau)^{-2}]$$

$$(10.1.21a) \qquad G_2(\tau) = 2\zeta(2) + \sum_{n=1}^{\infty} \sum_{m=-\infty}^{\infty} [(n\tau + m)^{-2}$$

$$+ (n\tau - m)^{-2}]$$

$$(10.1.21b) \qquad G_2(-1/\tau)/\tau^2 = 2\zeta(2) + \sum_{n=-\infty}^{\infty} \sum_{m=1}^{\infty} [(m\tau + n)^{-2}$$

$$+ (m\tau - n)^{-2}].$$

We now can see a forbidden interchange in the order of summation! For instance, if we set $G_*(\tau) = G_2(-1/\tau)/\tau^2$, then $G_*(\tau)$ could not equal $G_2(\tau)$ because of a formal interchange of summations. Indeed, we shall note that $G_*(\tau + 1) \neq G_*(\tau)$.

We can now apply the Poisson summation formula (10.1.14c) to the sums in equations (10.1.21a, b). The first is left for Exercise 10.2; we consider only equation (10.21b) here. We start with the gamma-integral,

$$(10.1.22a) \quad (m\tau + n)^{-2} = -4\pi^2 \int_0^{\infty} [\exp 2\pi i(m\tau + n)t] t\, dt$$

since Im $\tau > 0$ and $m > 0$. Then putting together two such integrals and breaking the interval from (real) 0 to ∞ into subintervals of length one, we find

$$(10.1.22b) \quad (m\tau + n)^{-2} + (m\tau - n)^{-2} = -8\pi^2 \sum_{h=0}^{\infty} \int_0^1 f(t)(\cos 2\pi nt) \, dt$$

for

$$(10.1.22c) \quad f(t) = \sum_{m=1}^{\infty} \sum_{h=0}^{\infty} [\exp 2\pi i m\tau(t + h)](t + h)$$

or more simply

$$(10.1.22d) \quad f(t) = \sum_{h=0}^{\infty} q^{(t+h)}(t + h)/(1 - q^{(t+h)}).$$

We note that $f(0)$ and $f(1)$ differ by a single term, which must be evaluated as a limit (as $t \to 0$):

$$(10.1.23a) \quad f(1) = \sum_{h=1}^{\infty} q^h/(1 - q^h)$$

$$(10.1.23b) \quad f(0) = f(1) + \lim_{t \to 0} q^t t/(1 - q^t) = f(1) - 1/2\pi i\tau.$$

We recognize $f(1)$ as part of $G_2(\tau)$ and $1/2\pi i\tau$ as the difference caused by rearrangement. Thus, finally,

$$(10.1.24) \quad G_2(-1/\tau)/\tau^2 = G_2(\tau) - 2\pi i/\tau.$$

Thus, $G_2(\tau)$ is not a modular invariant but a function that will lead us to one. If we define $H_0(\tau)$ by the primitive

$$(10.1.25a) \quad H_0(\tau) = \int G_2(\tau) \, d\tau$$

then $H_0(-1/\tau) = H_0(\tau) - 2\pi i \log \tau + \text{constant}$. So we set

$$(10.1.25b) \quad H(\tau) = (1/\tau) \exp H_0(\tau)/\pi i$$

and we find

$$(10.1.25c) \quad H(-1/\tau) = H(\tau) \text{ constant.}$$

There is a more traditional form for $H(\tau)$ with a natural choice of constant. By integrating the value in equation (10.1.19c) for $G_2(\tau)$ in equation (10.1.25b) (note $d\tau = dq/2\pi iq$), we find

$$(10.1.26a) \quad H(\tau) = \eta(\tau)^{-4}/\tau$$

where $\eta(\tau)$ is the Dedekind eta-function,

$$(10.1.26b) \qquad \eta(\tau) = q^{1/24} \prod_{h=1}^{\infty} (1 - q^h).$$

Thus, from equation (10.1.25c), $\eta(-1/\tau) = \eta(\tau)\sqrt{\tau} \cdot$ constant. [The constant is determined by the use of i, $\eta(-1/i) = \eta(i)$.] Thus,

$$(10.1.27) \quad \eta(-1/\tau) = \sqrt{\tau/i}\eta(\tau), \qquad \eta(\tau + 1) = \eta(\tau) \exp 2\pi i/24.$$

We are not directly interested in $\eta(\tau)$ but rather in

$$(10.1.28) \qquad \begin{aligned} \Delta(\tau) &= \eta(\tau)^{24} = q \prod_{h=1}^{\infty} (1 - q^h)^{24} \\ &= q - 24q^2 + 252q^3 - 1472q^4 + 4380q^5 - \cdots . \end{aligned}$$

This is precisely the function defined in equation (10.0.8). Indeed, we shall show in Section 10.3 that

$$(10.1.29) \qquad\qquad 1728\Delta(\tau) = E_4^3 - E_6^2.$$

We shall indeed deduce it from the following properties [see equations (10.1.27)]:

$$(10.1.30) \qquad \Delta(\tau) = \Delta(\tau + 1), \qquad \Delta(\tau) = \Delta(-1/\tau)/\tau^{12}.$$

10.2. Special matrix and transformation groups

We consider groups associated with the modular function $j(\tau)$ or $J(\tau)$. The action on $j(\tau)$ is given by a linear fractional transformation

$$(10.2.1) \qquad \begin{aligned} &\tau \rightarrow A(\tau), \qquad A(\tau) = (a\tau + b)/(c\tau + d) \\ &a, b, c, d \in \mathbf{Q} \quad \text{and} \quad \det A = ad - bc > 0. \end{aligned}$$

The determinant condition is required so that if Im $\tau > 0$, then

$$(10.2.2) \qquad \text{Im } A(\tau) = \text{Im } \tau (ad - bc)/|c\tau + d|^2 > 0.$$

For purposes of computation, we think of A as a matrix, however (using the same symbol). Thus,

$$(10.2.3a) \quad A = \begin{pmatrix} a & b \\ c & d \end{pmatrix}; \qquad a, b, c, d \in \mathbf{Q}, \quad \det A = ad - bc > 0.$$

Clearly, any matrix $B = rA$ for $r \in \mathbf{Q}^*$ has the same action on τ; that is, $B(\tau) = A(\tau)$. We eliminate this ambiguity only in a qualified sense by requiring that all matrices be integral and primitive (i.e., with no common divisor of the matrix elements). Thus the set arises:

(10.2.3b) $G^* = \{A : a, b, c, d \in \mathbf{Z}, \quad \gcd(a, b, c, d) = 1, \quad \det A > 0\}.$

Because A and $-A$ generate the same transformation $\tau \to (\pm A)(\tau)$, there is a two-to-one correspondence between matrix groups (denoted typically with asterisks, e.g., Γ^*) and transformation groups (e.g., Γ). In any case, the transformations have the same det A. Thus, we have a group of transformations G with identity 1, namely, $\tau \to \tau$ (corresponding to I and $-I$ both). We write typically

(10.2.3c) $G = \{\tau \to A(\tau) : A \in G^*\}.$

10.2.4. Definition. *The "matrix" modular group Γ^* is the set of matrices [classically $SL_2(\mathbf{Z})$]:*

(10.2.4a) $\Gamma^* = \{A : A \in G^*, \quad \det A = 1\}.$

The "transformation" modular group Γ is the set of transformations [classically $PSL_2(\mathbf{Z})$]:

(10.2.4b) $\Gamma = \{\tau \to A(\tau) : A \in G^*, \quad \det A = 1\}.$

It is usually clear from context which "modular" group is required (and likewise for subgroups of Γ^* and Γ). They are related by

(10.2.4c) $\Gamma = \Gamma^*/(\pm I)$

(and likewise for subgroups of each).

10.2.5. Definition. *The "elementary" matrices $\{E\}$ of G^* are those matrices that perform the "standard" row operations (on the left) or column operations (on the right) on a matrix A (but preserve det A),*

$$(10.2.5a) \quad \{E\} = \left\{ \begin{pmatrix} 1 & m \\ 0 & 1 \end{pmatrix}, \begin{pmatrix} -1 & 0 \\ 0 & -1 \end{pmatrix}, \begin{pmatrix} 0 & 1 \\ -1 & 0 \end{pmatrix}, \begin{pmatrix} 1 & 0 \\ m & 1 \end{pmatrix} \right\}, \quad m \in \mathbf{Z}.$$

(The group generated by $\{E\}$ is sometimes also called "elementary.")

We can use elementary matrices to reduce matrices $A^* \in G^*$ to normal forms as follows (remembering that A is primitive):

10.2.6. Smith–Hermite–Kronecker theorem. *If $A \in G^*$, then there is a unique normal form of the following types:*

(10.2.6a) $EA = \begin{pmatrix} a & b \\ 0 & d \end{pmatrix}$, $ad = \det A$, $a > 0$,

$$d > 0,\ d - 1 \geq b \geq 0,\ (a, b, d) = 1$$

(10.2.6b) $AE' = \begin{pmatrix} a & 0 \\ c & d \end{pmatrix}$, $ad = \det A$, $a > 0$, $d > 0$,

$$a - 1 \geq c \geq 0,\ (a, c, d) = 1$$

(10.2.6c) $EAE' = \begin{pmatrix} a & 0 \\ 0 & 1 \end{pmatrix}$, $a = \det A$

for matrices E, E' in the group generated by elementary matrices.

10.2.7. Corollary. *If A, $B \in G^*$ and $\det A = \det B$, then $EA = BE'$ for matrices E and E' as shown earlier. (Recall that all matrices are primitive.)*

10.2.8. Corollary. *The modular group Γ^* is precisely the group generated by the elementary matrices of Definition 10.2.5. The following are three traditionally used generators (any two suffice):*

(10.2.8a) $\tau \to \tau + 1 = U(\tau) \leftrightarrow \pm U$, $U = \begin{pmatrix} 1 & 1 \\ 0 & 1 \end{pmatrix}$

(10.2.8b) $\tau \to -1/\tau = T(\tau) \leftrightarrow \pm T$, $T = \begin{pmatrix} 0 & -1 \\ 1 & 0 \end{pmatrix}$

(10.2.8c) $\tau \to -1/(\tau + 1) = S(\tau) \leftrightarrow \pm S$, $S = TU = \begin{pmatrix} 0 & -1 \\ 1 & 1 \end{pmatrix}$.

Note that $S^2 = \begin{pmatrix} -1 & -1 \\ 1 & 0 \end{pmatrix}$ *and* $S^2(\tau) = -(\tau + 1)/\tau$. *Thus,*

(10.2.8d) $$S^3 = T^2 = -I.$$

10.2.9. Corollary. *The modular group has the presentation*

$$\Gamma = \langle S, T : S^3 = T^2 = 1 \rangle$$

where S and T now refer to transformations (see Exercises 10.3 and 10.4).

10.2.10. Definition. *The "principal" congruence subgroup of Γ^* (or of Γ) modulo n is defined as*

$$\Gamma^*(n) = \{M \in \Gamma,\quad M \equiv +I \bmod n\}$$
$$\Gamma(n) = \{\tau \to M(\tau),\quad M \equiv \pm I \bmod n\}.$$

A "congruence" subgroup H* (or H) is a subgroup of Γ* (or Γ) that contains some Γ*(n) [or Γ(n)]. Alternatively, a congruence matrix group is one whose elements are determined by the residue classes of the matrix elements modulo n. The minimum m for which a congruence subgroup H* ⊇ Γ*(m) is called the "conductor" of H* (and likewise of H). (Here two matrices are called "congruent" when the corresponding elements are.)

10.2.11. Theorem. *The index $|Γ^*/Γ^*(n)| = μ^*(n)$ and $|Γ/Γ(n)| = μ(n)$, where*

(10.2.11a) $μ^*(n) = n^3 \prod_{p|n} (1 - 1/p^2)$

(10.2.11b) $μ(n) = μ^*(n)/2$ *if $n > 2$* $[μ(n) = μ^*(n)$ *if $n = 1, 2$]*.

The proof follows the next steps (detailed in Exercise 10.5).

10.2.12. Lemma. $Γ^*/Γ^*(n) = SL_2(\mathbf{Z}/n\mathbf{Z})$.

10.2.13. Lemma. $SL_2(\mathbf{Z}/nm\mathbf{Z}) = SL_2(\mathbf{Z}/n\mathbf{Z}) \times SL_2(\mathbf{Z}/m\mathbf{Z})$, $(n, m) = 1$.

10.2.14. Lemma. $|SL_2(\mathbf{Z}/p^t\mathbf{Z})| = μ^*(p^t)$,   *p prime,* $t \geq 1$.

The quotient groups $Γ/Γ(n)$ for small n lead to classically known groups, symmetric $S(m)$ or alternating $A(m)$; for example, $Γ/Γ(2) = S(3)$, $Γ/Γ(3) = A(4)$, $Γ/Γ(4) = S(4)$, $Γ/Γ(5) = A(5)$, and so on. Recall that $|S(m)| = m!$ and $|A(m)| = m!/2$. We return to these groups in Chapter 11.

10.2.15. Definition. *The modular subgroup that belongs to a matrix $\mathbf{A} \in G^*$ is defined as a matrix or transformation group as follows:*

(10.2.15a) $Γ^*[A] = \{M \in Γ^* : AMA^{-1} \in Γ^*\}$

(10.2.15b) $Γ[A] = \{τ \to M(τ) : M \in Γ^*[A]\}$.

Note that A^{-1} is not in G^* if det $A > 1$, but $Γ^*[A]$ contains $\pm I$ so there is a two-to-one correspondence between $Γ^*[A]$ and $Γ[A]$.

10.2.16. Lemma. *For A primitive (as is usual), $Γ^*[A] \supseteq Γ^*$ (det A) and $Γ[A] \supseteq Γ(det A)$. Indeed, det A is the conductor in each case.*

Proof. Let $M \equiv I$ mod n and set $A^* = nA^{-1}$ for $n = \det A$, so $A^* \in G^*$ also. If $B = AMA^*$, then $B \equiv \pm AA^* \equiv nAA^{-1} \equiv 0$ mod n. Thus, $B/n = AMA^{-1} \in G^*$, whereas $\det B/n = \det AMA^{-1} = \det M = 1$. The conductor property is referred to in Exercise 10.6.   □

10.2.17. Lemma. *All $\Gamma^*[A]$ are isomorphic for a given $\det A$.*

Proof. If $B = E'AE$ for $E, E' \in \Gamma^*$ and $A \in G^*$, then $\Gamma[B] = E\Gamma[A]E^{-1}$ $(\Gamma[E'B] = \Gamma[B])$.   □

10.2.18. Definition. *Special cases of $\Gamma^*[A]$ or $\Gamma[A]$ are*

$$(10.2.18a) \quad \Gamma_0^*(n) = \Gamma^* \left[\begin{pmatrix} n & 0 \\ 0 & 1 \end{pmatrix} \right] = \left\{ \begin{pmatrix} a & b \\ c & d \end{pmatrix} \in \Gamma^*, \quad c \equiv 0 \bmod n \right\}$$

$$(10.2.18b) \quad \Gamma^{*0}(n) = \Gamma^* \left[\begin{pmatrix} 1 & 0 \\ 0 & n \end{pmatrix} \right] = \left\{ \begin{pmatrix} a & b \\ c & d \end{pmatrix} \in \Gamma^*, \quad b \equiv 0 \bmod n \right\}$$

with $\Gamma_0(n)$ and $\Gamma^0(n)$ consisting of the corresponding transformation groups.

10.2.19. Theorem. *The index $|\Gamma^*/\Gamma^*[A]| = |\Gamma/\Gamma[A]| = \psi(\det A)$, where $\psi(n) = n \sum_{p|n} (1 + 1/p)$. (See Exercise 10.7.)*

10.3. Modular fields and subfields

The matrix (and transformation) groups belong in a special way to function fields over $C(z)$ where now z stands for the transcendental variable $j(\tau)$. Hence, such function fields must correspond to Riemann surfaces. The important point is that the Riemann surfaces are not abstract manifolds but specific imbeddings in the upper half-plane:

$$(10.3.1) \qquad\qquad H = \{\tau : \operatorname{Im} \tau > 0 \}$$

built around fixed points of Γ. These will correspond to important algebraic values of $j(\tau)$, and so number theory and complex analysis are amalgamated.

10.3.2. Definition. *For a group G acting on H, namely,*

$$G = \{\tau \to g(\tau)\}$$

(where g maps H biuniquely onto H), the open region $\mathcal{D}$ (or its closure $\overline{\mathcal{D}}$) is called a "fundamental" domain when every point $\tau \in H$ corre-

sponds to at least one image $g(\tau)$ in $\overline{\mathcal{D}}$. Furthermore, this is unique if $g(\tau)$ is in $\mathcal{D}$; that is, if $g_1(\tau) = g_2(\tau)$ for some τ and both lie in $\mathcal{D}$, then the functions $g_1(\tau)$ and $g_2(\tau)$ are identical. Otherwise expressed, if $\tau \in \mathcal{D}$ and $g(\tau) \in \mathcal{D}$, then g is the identity function. We write $\mathcal{D} = fd(G)$.

There are rather broad theorems to reassure us that if H/G has a discrete quotient topology, then a fundamental domain for G must exist. Also, from the continuity of the functions $g(\tau)$, there must be functions that identify segments of the boundary (and possibly isolated fixed points). We merely use special properties of the modular group.

10.3.3. Theorem. *The modular group Γ has a fundamental domain $\mathcal{D}$ that can be described (among a number of ways) as follows:*

(10.3.3a) $$-\tfrac{1}{2} < Re\, \tau < \tfrac{1}{2}$$
(10.3.3b) $$1 < |\,\tau\,|.$$

The boundary segments map into one another by the following transformations:

$\tau \to U(\tau) = \tau + 1$ *(matching vertical segments)*
$\tau \to T(\tau) = -1/\tau$ *(matching circular arcs on both sides of i).*

The fixed points of $\overline{\mathcal{D}}$ (mapped into themselves by G) are as follows:

i *is a fixed point of* $T(\tau) = -1/\tau$
$\rho = (-1 + \sqrt{-3})/2$ *is a fixed point of* $S(\tau) = -1/(\tau + 1)$
    and $S^2(\tau) = -(\tau + 1)/\tau$
$\rho^* = (1 + \sqrt{-3})/2$ *is a fixed point of* $TS^2T(\tau) = -1/(\tau - 1)$
    and $TST(\tau) = (\tau - 1)/\tau.$

(See Figure 10.1.)

First proof. To understand where the inequalities (10.3.3a, b) arise, consider the lattice $\mathcal{L} = [2\omega_1, 2\omega_2]$. To find an invariant basis for the lattice, we can demand that one basis vector be the shortest vector in the lattice. Such a vector must exist if we assume that $\omega_1/\omega_2 = \tau \in H$, although there might be two, four, or six of them (see Exercise 10.9). Regardless, the second basis vector can be chosen as the shortest vector that is not collinear with the first. Then if ω_1 is the first, $\tau\omega_1$ is the second, for $\tau \in \overline{\mathcal{D}}$. $\square$

10.3.4. Lemma. *Let $|Re\,\tau| \le \tfrac{1}{2}$ and $Im\,\tau \le \tfrac{1}{2}$; then $T(\tau)$ satisfies $Im\, T(\tau) \ge 2\,Im\,\tau$.*

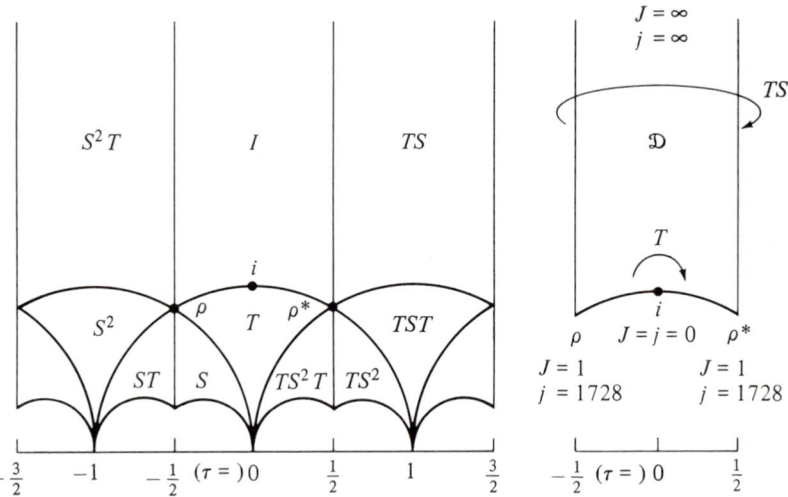

Figure 10.1. Fundamental domain $\mathcal{D}$ for modular group (right), showing action of generators (left).

Proof. Note that Im $T(\tau) = $ Im $\tau/|\tau|^2$, and by hypothesis, $|\tau|^2 < (\frac{1}{2})^2 + (\frac{1}{2})^2$.   □

Second proof of Theorem 10.3.3. Let $\{x\}$ be the integer closest to the real number x, so

$$-\tfrac{1}{2} \le x - \{x\} < \tfrac{1}{2}.$$

Then if τ is given in the upper half-plane H, set

$$\tau_1 = \tau - (\text{Re } \tau).$$

Either $\tau_1 \in \overline{\mathcal{D}}$ or Im $\tau_1 < 1$ (or both), so we can set up the continued fraction sequence

$$\tau_{n+1} = \{\text{Re } 1/\tau_n\} - 1/\tau_n \qquad (n \ge 1)$$

stopping as soon as Im $\tau_n > \frac{1}{2}$. Set $\tau_{n+1} = V(\tau_n)$ where $V = S^2$, T, or TST (see Figure 10.1), so $\tau_{n+1} \in \overline{\mathcal{D}}$. But, by Lemma 10.3.4, we can have Im $\tau_n \le \frac{1}{2}$ for at most $\log_2 (1/\text{Im } \tau)$ values of n. Thus, an effective bound exists for the number of continued fraction inversions that are required to reduce τ to its image in $\overline{\mathcal{D}}$.   □

10.3.5. Theorem. *The modular invariant $j = j(\tau)$ maps the fd(Γ) onto the j-plane so that every point is covered exactly once (including the* *boundary under the identifications of Theorem (10.3.3). The mapping is bianalytic in the "natural" $\tau \leftrightarrow j$ local coordinates except as follows:*

$$\tau = \rho, \quad j = const(\tau - \rho)^3 \quad \text{(likewise for } \rho\text{*)}$$
$$\tau = i, \quad j - 1728 = const(\tau - i)^2$$
$$\tau = i\infty, \quad j = const/q \quad (q = exp\, 2\pi i\tau).$$

(Here "const" refers to nonzero constants.)

Proof. The image $j(\overline{\mathcal{D}})$ by boundary identifications must cover the finite j-plane a finite number of times (actually once from Theorem 10.3.3). The other local behavior is explained by the geometry of the angles at the various fixed points. (The neighborhood of ρ and ρ^* is effectively 120° for exponent 3 and the neighborhood of i is 180° for exponent 2.) $\square$

10.3.6. Theorem. *Let Γ_I be a subgroup of Γ of finite index h, with left cosets $\{S_1, \ldots, S_h\}$ (e.g., $S_1 = 1$):*

$$\Gamma = S_1\Gamma_I \cup \cdots \cup S_h\Gamma_I.$$

Then the fundamental domain of Γ_I is $\mathcal{D}_I$, where

$$\mathcal{D}_I = S_1^{-1}(\mathcal{D}) \cup \cdots \cup S_h^{-1}(\mathcal{D}), \quad \mathcal{D} = fd(\Gamma).$$

For convenience, the coset representatives S_j are chosen successively so that $\overline{\mathcal{D}}_1$ will be connected.

10.3.7. Definition. *The "cusp" points of $fd(\Gamma_I) = \mathcal{D}_I$ are the distinct images of ∞ to within elements of Γ_I [effectively, the distinct images generated by $S_j(\infty)$ for $j = 1, \ldots, h$].*

10.3.8. Lemma. *The fixed points of Γ are merely images of i and ρ under Γ; hence, the fixed points of any Γ_I are merely some subset of these points (possibly empty). (See Exercise 10.10.)*

10.3.9. Definition. *The compact Riemann surface $\mathcal{M}(\Gamma_I)$ is the set of neighborhoods in $\overline{\mathcal{D}}_I$ identified under Γ_I with a local analytic structure defined by $j(\tau)$ according to Theorem 10.3.3, with cusp points and fixed points (if any) as noted.*

10.3.10. Definition. *The modular function field $\mathcal{F}(\Gamma_I)$ is the function field over $\mathcal{M}(\Gamma_I)$. Its genus is called the "genus of Γ_I."*

10.3.11. Remark. Clearly, $\mathscr{F}(\Gamma) = \mathbf{C}(j)$ for $j = j(\tau)$ and, generally, $\mathscr{F}(\Gamma_1) = \mathbf{C}(j, y)$, where y satisfies an algebraic equation over $\mathbf{C}(j)$ of degree $|\Gamma/\Gamma_1|$ (since this degree is the number of y for each j).

10.3.12. Definition. *A "differential" of degree m on $\mathcal{M}(\Gamma_1)$ is a formal expression $f(\tau)(d\tau)^m$ that is invariant under Γ_1, or*

(10.3.12a) $$f(S(\tau)(dS(\tau))^m = f(\tau)(d\tau)^m$$

for all $S \in \Gamma_1$. [Alternatively, $f(\tau)/j'(\tau)^m$ is invariant.] If $S = \begin{pmatrix} a & b \\ c & d \end{pmatrix}$, det $S = 1$,

(10.3.12b) $$\begin{aligned} S(\tau) &= (a\tau + b)/(c\tau + d), \\ dS(\tau)/d\tau &= (c\tau + d)^{-2}, \ f(S(\tau))(c\tau + d)^{-2m} = f(\tau). \end{aligned}$$

We call such an $f(\tau)$ a "modular form" of degree m if $f(\tau)$ is analytic (has no poles) in the whole upper half-plane and is bounded at cusp points. The last condition means that for every cusp point $S_j(\infty)$, we have the condition that $f(S_j(\tau))$ is bounded as $\tau \to i\infty$. We say that $f(\tau)$ has a "cusp" (for $i\infty$) if it approaches zero as $\tau \to i\infty$, and such an $f(\tau)$ is called a "cusp form."

10.3.13. Remark. For $\mathcal{M}(\Gamma)$, the Eisenstein series $G_{2k}(\tau)$ [or $E_{2k}(\tau)$] are modular forms of degree k. [Note in equation (10.1.3) that the invariance is shown for the generators of Γ, and this suffices.] The justification for the term "form" lies in the analogy with $G_{2k}(2\omega_1, 2\omega_2)$, which is homogeneous (of degree $-2k$). The discriminant function $\Delta(\tau)$ is, however, a cusp form because its q-series begins with a vanishing term. Also note that modular forms are a graded ring and a vector space over $\mathbf{C}$.

10.3.14. Theorem. *The space $L_m(\Gamma_1)$ of modular forms on $\mathcal{M}(\Gamma_1)$ of dimension m and $K_m(\Gamma_1)$ the space of cusp forms are of finite dimension over $\mathbf{C}$.*

Proof. We are looking for forms $f(\tau)$ for which, by definition, $g = f/j'^m$ lies in the function field $\mathscr{F}(\Gamma_1)$. Now by the Riemann–Roch theorem (see Chapter 5), g has poles limited (finitely) by zeros of $j'(\tau)$ at the branch points and by the estimate $|g| < \text{const}/q^m$ at $i\infty$ (and similarly for other cusp points). $\qquad\square$

10.3.15. Theorem. *For the modular group* Γ,

$$dim_C L_m(\Gamma) = [2m/3] + [m/2] - m + 1$$

$$dim_C L_m(\Gamma) = \begin{cases} [m/6] & \text{if } m \equiv 1 \bmod 6 \\ [m/6] + 1 & \text{otherwise} \end{cases}$$

$$dim_C K_m(\Gamma) = dim_C L_m(\Gamma) - 1.$$

Proof. If $f \in L_m(\Gamma)$, then $f = j'^m R(j)$, where R is a rational function (see Remark 10.3.11). Since f is analytic and bounded at $i\infty$, R is constrained to be $P(j)j^{-[2m/3]}(j - 1728)^{-[m/2]}$ with the degree of the polynomial $P(j) \le [m/2] + [2m/3] - m$ because j and j' have the same order of magnitude at $i\infty$. (This follows from Theorem 10.3.5.) $\square$

10.3.16. Corollary. *In the preceding result, $dim_C L_m(\Gamma) = 1$ for $m = 2$, 3, 4, 5, and $dim_C K_m(\Gamma) = 1$ for $m = 6$. Thus, $L_m(\Gamma)$ is spanned by the following:*

$$
\begin{aligned}
m &= 2, & E_4(\tau) \\
m &= 3, & E_6(\tau) \\
m &= 4, & E_8(\tau) \, [= E_4(\tau)^2] \\
m &= 5, & E_{10}(\tau) \, [= E_4(\tau)E_6(\tau)] \\
m &= 6, & E_4(\tau)^3, \, E_6^2(\tau).
\end{aligned}
$$

Likewise, $K_6(\Gamma)$ is spanned by $1728\Delta(\tau) = E_4(\tau)^3 - E_6(\tau)^2$. (Note that the above identities are of special interest.)

10.3.17. Remark on Galois theory and class field analogies. Although we deal with only special groups $\Gamma[A]$ and the corresponding modular function fields, we can give some thought to the aggregate of such groups and fields. First of all, as in the case of number fields, the normal closures are taken, so Galois theory has meaning. Accordingly, let Γ_0 be a normal subgroup of Γ of finite index h, with (right or left) cosets S_i, where $i = 1, \ldots, h$. Let $\mathcal{F}[\Gamma_0]$ be the corresponding (normal) function field over $\mathcal{F}[\Gamma]$ $[= C(j)]$. It is clear that if $f(\tau)$ is a function in $\mathcal{F}[\Gamma_0]$, then the set of $f(S_i(\tau))$ is the conjugates over $C(j)$ and

$$\text{Gal } \mathcal{F}[\Gamma_0]/C(j) = \Gamma/\Gamma_0.$$

Clearly, such Γ_0 (and the corresponding $\mathcal{F}[\Gamma_0]$) form inclusion and intersection lattices within Γ [or correspondingly over $C(j)$]. Now, within this lattice lies a sublattice of congruence subgroups of Γ [and the so-called congruence modular function fields over $C(j)$]. (Unfortunately, there are also noncongruence subgroups!) This smaller congruence lattice of fields is like the (abelian) class fields over k (lying within the larger lattice of normal fields over k). The conductor plays the same role: Every congruence field lies in a field $\mathscr{F}[\Gamma(n)]$ of a principal congruence group of conductor n, just as every class field lies in a ray class field of conductor (say) f. It is still not clear which algebraic property (if any) in the congruence subgroup lattice plays the same role as the abelian property of Galois theory.

10.4. Modular equations

We are concerned with the function field $\mathbf{C}(j(n\tau), j(\tau))$ as a finite extension over $\mathbf{C}(j(\tau))$. It is necessary to simultaneously consider all function fields $\mathbf{C}(j(A\tau), j(\tau))$, where $\det A = n$, because these are all conjugates. [We now abbreviate $A(\tau)$ as A for convenience.] A more precise result is the following:

10.4.1. Theorem. *Let $A \in G^*$ (A primitive) and let det $A = n$. For such A we define the functions*

$$(10.4.1a) \quad j_A(\tau) = j(A\tau) = j((a\tau + b)/(c\tau + d))$$

$$A = \begin{pmatrix} a\ b \\ c\ d \end{pmatrix}, \quad ad - bc = n, \quad gcd(a, b, c, d) = 1.$$

It is possible to find a set of

$$\psi(n) = n \prod_{p \mid n} (1 + 1/p)$$

reduced matrices [see equation (10.2.6a)], namely,

$$(10.4.1b) \quad A_r = \begin{pmatrix} a\ b \\ 0\ d \end{pmatrix}, \quad ad = n, \quad 0 \le b \le d - 1$$

so that $j_A(\tau)$ is one of $\psi(n)$ conjugates $j(A_r\tau)$. Each $j_A(\tau)$ is invariant under the modular subgroup $\Gamma[A]$ [of index $\psi(n)$] corresponding to a field $\mathscr{F}(\Gamma[A]) = \mathbf{C}(t, j)$ of degree $\psi(n)$ over $\mathscr{F}(\Gamma) = \mathbf{C}(j)$, where we use the symbols

$$(10.4.1c) \quad t = j_A(\tau), \quad j = j(\tau).$$

Proof. Note that $j_A(\tau) = j_A(M\tau)$ when $j(A\tau) = j(AM\tau)$ or when $M'A = AM$ for $M, M' \in \Gamma$ (by the invariance of j under Γ). Hence, $j_A(\tau)$ is invariant when $M' = AMA^{-1} \in \Gamma$ for $M \in \Gamma[A]$. Clearly, $j_A(\tau) = j(M'A\tau)$, which leads to equations (10.4.1b). The field $\mathbf{C}(t, j)$ is of degree $\psi(n)$ over $\mathbf{C}(j)$ (and no lower) because all reduced matrices are accessible as M varies over Γ, with $j_A(M\tau) = j(MA\tau)$. $\qquad\square$

10.4.2. Illustrations. When $n \, (= p)$ is prime and $\psi(n) = p + 1$, then

$$(10.4.2a) \quad j(\tau/p), j((\tau + 1)/p), \ldots, j((\tau + p - 1)/p), j(p\tau)$$

are the conjugates of $j(p\tau)$. They are renumbered for convenience as

$$(10.4.2b) \quad j_b = j((\tau + b)/p), \qquad 0 \le b \le p - 1, \qquad j_p = j(p\tau)$$

corresponding to

$$(10.4.2c) \quad A_b = \begin{pmatrix} 1 & b \\ 0 & p \end{pmatrix}, \qquad 0 \le b \le p - 1, \qquad A_p = \begin{pmatrix} p & 0 \\ 0 & 1 \end{pmatrix}.$$

For $n = 4$, the reduced matrices lead to the conjugate set

$$(10.4.2d) \quad j((\tau + b)/4), \qquad 0 \le b \le 3, \qquad j(\tau + \tfrac{1}{2}), \quad j(4\tau).$$

10.4.3. Theorem. *The "modular equation" for $j(A\tau)$ is*

$$(10.4.3a) \qquad J_n(t, j) = \prod_{\psi(n)} [t - j((a\tau + b)/d))] = 0$$

for the $\psi(n)$ reduced matrices (10.4.1b). It is a polynomial over $\mathbf{C}$ in t and j of degree $\psi(n)$ in each variable; indeed, for $n > 1$,

$$(10.4.3b) \qquad J_n(t, j) = J_n(j, t).$$

Proof. All but the symmetry follows from the fact that the matrices (10.4.1b) give a complete set of conjugates. To complete the proof, note by equation (10.4.3a) that $J(\tau/n) = t$ is also a root. Thus, $J_n(j(\tau/n), j(\tau)) = 0$. With $n\tau$ replacing τ, we find $J_n(j, t) = 0$. In short,

$$J_n(t, j) = 0 \Rightarrow J_n(j, t) = 0.$$

Because J_n is irreducible in t,

$$J_n(j, t) = cJ_n(t, j)$$

for $c = c(t, j)$, a polynomial in t and j. By iteration,

$$J_n(j, t) = c(t, j)c(j, t)J_n(j, t).$$

Thus, c is a constant whose square is 1. We exclude $c = -1$ because if $J_n(t, j) = -J_n(j, t)$, it follows that $J_n(t, t) = 0$. Then $J_n(t, j)$ could no longer be irreducible because it would be divisible by $(t - j)$. $\qquad\square$

10.4.4. Lemma. *If $f(\tau)$ is a modular form of degree m and A is again a primitive matrix acting on τ, that is,*

$$A\tau = (a\tau + b)/(c\tau + d), \qquad ad - bc = n$$

then we define

$$(10.4.4a) \qquad f_A(\tau) = n^{2m}(c\tau + d)^{-2m}f(A\tau)/f(\tau).$$

We assert that $f_A(\tau)$ is a modular function belonging to the subgroup $\Gamma[A]$ that determines the same field as $j(A\tau)$ $(= t)$, so

$$(10.4.4b) \qquad f_A(\tau) \in \mathbf{C}(j, t).$$

Proof. We note for some specially chosen rational function $R(j)$,

$$(10.4.5a) \qquad f(\tau) = j'(\tau)^m R(j)$$
$$(10.4.5b) \qquad f(A\tau) = j'(A\tau)^m R(t).$$

We see that $j'(A\tau) = (dj(A\tau)/d\tau)/(d(A\tau)/d\tau)$, so

$$(10.4.5c) \qquad j'(A\tau) = (dt/d\tau)((c\tau + d)^2/n).$$

Thus, we have an equivalent formula

$$(10.4.5d) \qquad f_A(\tau) = n^m(dt/dj)^m R(t)/R(j).$$

This function clearly belongs to $\mathbf{C}(j, t)$ because dt/dj also does. [Recall that $dt/dj = -(\partial J_m/\partial j)/(\partial J_m/\partial t)$ by implicit differentiation.] $\qquad \square$

A particular case of importance is for $\Delta(\tau)$, given by

$$(10.4.6) \qquad \Delta(\tau) = (j'/2\pi i)^6 j^{-4}(j - 1728)^{-3}$$

$$= (E_4^3 - E_6^2)/1728 = q \prod_{h=1}^{\infty} (1 - q^h)^{24}.$$

We therefore define

$$(10.4.6a) \qquad \Delta_A(\tau) = n^{12}(c\tau + d)^{-12} \Delta(A\tau)/\Delta(\tau).$$

The function $\Delta(\tau)$ has the special property of not vanishing in the upper half-plane. For $n = p$ (prime), it has the $(p + 1)$ conjugates [see (10.4.2c)],

$$(10.4.6b) \quad \begin{cases} \Delta_b(\tau) = \Delta((\tau + b)/p)/\Delta(\tau), & 0 \le b \le p - 1 \\ \Delta_p(\tau) = p^{12} \Delta(p\tau)/\Delta(\tau). \end{cases}$$

Accordingly, $\Delta_b(\tau) \in \mathbf{C}(j_b(\tau), j(\tau))$.

10.4.7. Theorem. *The "multiplier equation" is defined by*

(10.4.7a)
$$\Phi(t, j) = \prod_{\psi(n)} (t - \Delta_A(\tau))$$

for $\psi(n)$ reduced matrices A [see (10.4.1b)]. Here $\Phi(t, j)$ is a polynomial in both j and t and its constant term in t, $\Phi(0, j)$, is a constant (in $\mathbf{C}$).

Proof. Because $\Delta(\tau)$ is nonsingular (no poles), the symmetric functions of Δ_A, which must lie in $\mathbf{C}(j)$, must also be polynomials (in $\mathbf{C}[j]$). The constant term, moreover, is a product of Δ_A and hence a nonvanishing polynomial, which, of course, is a constant in $\mathbf{C}$. □

To go further and discuss the integrality of the coefficients in the modular equation $J_n(t, j)$ and the multiplier equation $\Phi_n(t, j)$ we need the (power series) q-expansions.

10.4.8. Theorem on q-expansions (Hasse). *If a polynomial $P(j)$ has an expansion in q with coefficients in a ring R, then $P(j) \in R[j]$. (This also holds for a ring without unity, such as an ideal.)*

Proof (by induction). If the highest-order term (most negative power) in the q-expansion of $P(j)$ is a/q^n, then we replace $P(j)$ by $P(j) - aj^n$. Ultimately, $P(j)$ is reduced to a constant. □

10.4.9. Lemma. *The following are the q-expansions when $A = \begin{pmatrix} a & b \\ 0 & d \end{pmatrix}$:*

(10.4.9a) $\quad j_A = q^{-a/d}\zeta_d^{-b}(a + s(q^{a/d}, \zeta_d^b))$

(10.4.9b) $\quad \Delta_A = a^{12}q^{-1+a/d}\zeta_d^b(1 + s(q^{1/d}, \zeta_d^b))$.

Here ζ_d is exp $2\pi i/d$ as usual, and s is a power series over $\mathbf{Z}$ in the indicated variables. The power series is the same for each fixed pair a, d. In particular, for $n = p$ (prime), there is a power series s_b or t_b defined in each case:

(10.4.9c)
$$\begin{cases} j_b = q^{-1/p}\zeta_p^{-b}(1 + s_b(q^{1/p}, \zeta_p^b)), & 0 \le b \le p - 1 \\ j_p = q^p(1 + s_b(q^p)) \end{cases}$$

(10.4.9d)
$$\begin{cases} \Delta_b = q^{-1+1/p}\zeta_p^b(1 + t_b(q^{1/p}, \zeta_p^b)), & 0 \le b \le p - 1 \\ \Delta_p = p^{12}q^{p-1}(1 + t_b(q^p, 1)). \end{cases}$$

10.4.10. Theorem. *The polynomials $J_n(t, j)$ and $\Phi_n(t, j)$ are monic in t and have integral coefficients.*

Proof. We use Theorem 10.4.8 together with knowledge that the numerical coefficients are symmetric in $\mathbf{Q}(\zeta_d)/\mathbf{Q}$. □

10.4.11. Corollary. *The constant term $\Phi_n(0, j)$ is $\Pi_{\psi(n)}a^{12}$. In particular, $\Phi_p(0, j) = p^{12}$.*

We introduce the notation

$$A \equiv B \text{ mod. } c$$

to mean that two q-expansions, for A and B, have corresponding coefficients that are algebraic integers differing by multiples of c. The q-expansions may include fractional powers of q.

10.4.12. Theorem. $J_p(t, j) \equiv (t^p - j)(t - j^p)$ *mod. p. This congruence also holds mod p (in $\mathbf{Z}[t, j]$).*

Proof. From equations (10.4.9c), for $0 \le b \le p - 1$,

$$j_b \equiv q^{-1/p}\zeta_p^{-b} + \cdots \equiv j_0 \text{ mod. } (1 - \zeta_p)$$

since $\zeta_p^b \equiv 1 \bmod (1 - \zeta_p)$ in the usual sense. Thus,

$$\prod_{b=0}^{p-1} (t - j_b) \equiv (t - j_0)^p \equiv t^p - j_0^p \equiv t^p - j \text{ mod. } (1 - \zeta_p).$$

We use Fermat's theorem [recalling that $(1 - \zeta_p)|p$]. Again,

(10.4.12a) $j_p \equiv j_p$ so $t - j_p \equiv t - j^p$ mod. p.

Finally, apply Theorem 10.4.8 to

$$J_p(t, j) \equiv (t^p - j)(t - j^p) \text{ mod. } (1 - \zeta_p).$$

Note that p is the smallest rational integer divisible by $(1 - \zeta_p)$. □

By a similar computation based on equations (10.4.9d) we reach the following theorem (see Exercise 10.16).

10.4.13. Theorem. $\Phi_p(t, j) \equiv t^{p+1} + g(j)t$ *mod. p for g(j) a polynomial in $\mathbf{Z}[j]$. The same holds mod p (in $\mathbf{Z}[t, j]$).*

We next have to evaluate $j(\tau)$ for imaginary quadratic surds. They all come as roots of $J_n(t, t) = 0$. Yet we need the multiplier equation (as well as the modular equation) to establish the Galois group action on the roots of $J_n(t, t) = 0$. The interplay of the modular equation, $J_n(t, j) = 0$, and the multiplier equation, $\Phi_n(t, j) = 0$, comes from the fact that the $\psi(n)$ conjugates $t = j_A(\tau)$ or $t = \Delta_A(\tau)$ generate the same field $\mathcal{F}_A = \mathbf{Q}(j, j_A) = \mathbf{Q}(j, \Delta_A)$ over $\mathbf{Q}(j)$. The extensions are not normal; their closure (not required here) is of degree $\mu(n) = |\Gamma/\Gamma(n)|$.

10.4.14. Theorem. *Let $f_A \in \mathbf{Z}[j(\tau), j_A(\tau)]$. Then for fixed polynomials $a_m(j) \in \mathbf{Z}[j]$ (as A varies over canonical transformations of determinant n),*

(10.4.14a)
$$f_A \Phi'_n(\Delta_A, j) = \sum_{m=0}^{\psi(n)-1} a_m(j) \Delta_A^m$$

where $\Phi'_n(t, j)$ means $\partial \Phi_n/\partial t$ [not identically zero for any $t = \Delta_A(\tau)$].

Proof. From the construction of the different (6.1.12c) we see that the result holds, but only in the weaker form $a_m(j) \in \mathbf{Q}[j]$. To see the stronger result, note that f_A has a q-series of algebraic integers; so by symmetric functions [see equations (10.4.9a, b)], we see that the f_A could also be derived as solutions of the system

(10.4.14b) $\quad \sum_A f_A \Delta_A^m \in \mathbf{Z}[j], \quad m = 0, \ldots, \psi(n) - 1.$

Hence, $a_m(j) \in \mathbf{Z}[j]$. [Because each Δ_A is distinct, $\Phi'_n(\Delta_A, j) \neq 0$.] $\quad\square$

Exercises

10.1. Define the generating function $\Phi(u, x) = \sum_{n=0}^{\infty} u^n B_n(x)/n!$ Verify the properties $\Phi(0, x) = 1$, $\partial\Phi/\partial x = u\,\Phi(u, x)$ and $\int_0^\infty \Phi(u, x)\,dx = 1$ from equations (10.1.10). Deduce the value $\Phi(u, x) = u \exp ux/(\exp u - 1)$, and from this, deduce equation (10.1.12) as well as $u \cot u = \sum_0^\infty (-4u^2)^m B_{2m}/(2m)!$

10.2. Use the Poisson summation formula (10.1.5) to deduce equation (10.1.7) by starting with the analog of (10.1.22a),

$$(m\tau + n)^{-2k} = (2\pi i)^k \int_0^\infty [\exp 2\pi i(m\tau + n)t]t^{2k-1}\,dt.$$

10.3. Verify that the elementary matrices (10.2.5a) are generated by any two of S, T, and U.

10.4. Verify the presentation in Corollary 10.2.9; that is, verify that the matrix $W = \Pi\, S^{a_i} T^{b_i}$ can be $\pm I$ only trivially. (This means that if $W = I$, the powers of S and T can be grouped into factors of S^3 and T^2.) *Hint:* Use the fact that $-TS$ and $-TS^2$ are matrices with all positive elements; hence, a nontrivial W cannot be I or $-I$.

10.5. Complete the proof of Theorem 10.2.11 according to Lemmas 10.2.12, 10.2.13, and 10.2.14. *Hint for Lemma 10.2.14:* For $SL_2(\mathbf{Z}/p\mathbf{Z})$, note that the entries a, b, c, d modulo p fall into four categories of matrices (with $ad - bc \equiv 1 \bmod p$) as follows:

$$(a, p) = 1,\ ad \equiv 1,\ b \equiv 0 \rightarrow p(p - 1) \text{ matrices mod } p$$
$$(a, p) = 1,\ ad \equiv 1,\ b \not\equiv 0,\ c \equiv 0 \rightarrow (p - 1)^2 \text{ matrices mod } p$$
$$(a, p) = 1,\ ad \not\equiv 1,\ b \equiv 0 \rightarrow (p - 1)^3 \text{ matrices mod } p$$
$$a \equiv 0,\ b \not\equiv 0 \rightarrow p(p - 1) \text{ matrices mod } p.$$

[The total is $\mu^*(p)$.] We verify inductively that $|\Gamma^*/\Gamma^*(p^{t+1})| = p^3 |\Gamma^*/\Gamma^*(p^t)|$ by the substitution $a \rightarrow a + a_0 p^t$, and so on. This boosts the matrices from the modulus p^t to p^{t+1}. Explain the factor of $\frac{1}{2}$ present in $\mu(n)$ for $n > 2$.

10.6. Prove the conductor property of Lemma 10.2.16. *Hint:* Use Lemma 10.2.17 to simplify the choice of A.

10.7. Prove Theorem 10.2.19 by referring to the counting method of Exercise 10.5. Alternatively, deduce $\psi(n)$ as the number of normal forms (10.2.6ab) directly from Lemma 10.2.17 and equation (10.2.6c).

10.8. Verify that the matrices for which $ab + cd + 3bc \equiv 0 \bmod 6$ form a congruence subgroup of Γ^* (and of Γ) with index 6 and conductor 6. (This group is the commutator subgroup of either modular group. It is generated by TS^2TS and $TSTS^2$.)

10.9. Verify that the fundamental domain in Theorem 10.3.3 agrees with Lagrange's criterion for the reduction of a positive definite quadratic form over $\mathbf{Z}$ (indeed, even over $\mathbf{R}$) as follows:

$$F = ax^2 + bxy + cy^2, \qquad a > 0,\quad d = b^2 - 4ac < 0$$
$$F = \mathrm{norm}\,(\omega), \qquad \omega = x\omega_1 + y\omega_2 \in [\omega_1, \omega_2]$$
$$\omega_1 = \sqrt{a}, \qquad \omega_2 = (b + \sqrt{d})/(2\sqrt{a}).$$

Note that $|\omega_1| = a$, $|\omega_2| = c$, and $\mathrm{Re}\,\omega_1/\omega_2 = b/a$. Thus, $-a \leq b \leq a \leq c$. Show that the ambiguities in the choice of the basis arise with the forms

$$x^2 + y^2 \quad \text{and} \quad x^2 + xy + y^2.$$

10.10. Show Lemma 10.3.8 by considering when a fixed point $\tau = (a\tau + b)/(c\tau + d)$ can be nonreal while $ad - bc = 1$. Can $\Gamma(n)$ have any fixed points?

10.11. Write $G_4(\tau)$, $G_6(\tau)$, and $\Delta(\tau)$ explicitly in terms of $j(\tau)$ and $j'(\tau)$. Find identities that connect E_{12}, E_4, and E_6 from $L_6(\Gamma)$.

10.12. Prove directly that $E_4^3 - E_6^2$ has no zeros in Im $\tau > 0$ (see Exercise 10.11) and that all its coefficients are divisible by 1728 (without using the q-expansion of Δ). *Hint:* Note that $a^3 \equiv a^5$; hence $\sigma_3(m) \equiv \sigma_5(m)$ mod 12. Thus, the series $3 \cdot 240 \sum \sigma_3(m)q^m + 2 \cdot 504 \sum \sigma_5(m)q^m$ has coefficients divisible by 1728.

10.13. Verify that the group defined by $\Gamma[n] = \cap \Gamma[M]$ over all M of det $M = n$ is given by $\{M : M \equiv aI, a^2 \equiv 1 \bmod n\}$. Note that $\Gamma[n] \neq \Gamma(n)$ if $\mathbf{Z}/n\mathbf{Z}^*$ is not cyclic (e.g., $n = 8$, 15, etc.). *Hint:* It suffices to try just these three matrices:

$$M = \left\{ \begin{pmatrix} n & 0 \\ 0 & 1 \end{pmatrix}, \begin{pmatrix} 1 & 0 \\ 0 & n \end{pmatrix}, \begin{pmatrix} 1 & 1-n \\ 1 & 1 \end{pmatrix} \right\}.$$

10.14. Show that the Galois closure of $\mathbf{C}(j(n\tau), j(\tau))$ over $\mathbf{C}(j(\tau))$ is not necessarily $\mathscr{F}(\Gamma(n))$ but a subfield $\mathscr{F}_*$ for which Gal $\mathscr{F}(\Gamma(n))/\mathscr{F}_* = (\mathbf{Z}/2\mathbf{Z})^{s-1}$, where s is the two-rank of $\mathbf{Z}/n\mathbf{Z}^*$.

10.15. Show directly from the q-(product) expansion that $\Delta(\tau/2)\Delta((\tau + 1)/2)\Delta(2\tau) = \Delta(\tau)^3$. Verify that this is a consequence of Corollary 10.4.11.

10.16. Prove Theorem 10.4.13 by noting from equations (10.4.9d) that $\Delta_0 \equiv \Delta_b$ mod. $(1 - \zeta_p)$ for $0 \le b \le p - 1$ and $\Delta_p \equiv 0$; hence, $\Phi_p(t, j) \equiv (t - \Delta_0)^p t$.

Bibliographic note

Overall reference is to Deuring (1958), Weber (1891), Fueter (1924), Hasse (1927b, 1931), Ogg (1973), and Serre (1966), with Rankin (1969) for modular subgroups. The modular equation is treated in Birch (1973), Fricke and Klein (1890), and Yui (1978).

11

Class fields by modular functions

This chapter completes the proof of Weber's theorem, a result that is a unique accomplishment as far as it goes. It gives a specific answer: For a ring discriminant $d < 0$,

$$\text{RCF}\{d\} = \mathbf{Q}(\sqrt{d}, j((d + \sqrt{d})/2)).$$

The problem remains to evaluate j, but with the foreboding that all but the smallest values will be beyond human comprehension in magnitude! The nineteenth-century mathematician had no interest in the abstract concept of "effective computability" but had considerable interest and patience with manipulative and numerical computability. This was fortunate because some of the most beautiful theoretical results involving the modular equation came from efforts to reduce computations to human proportions. We refer to the results usually put under the heading of "Klein's icosahedron," results that are of such profound importance that they are generally detached from their source and stand alone on their intrinsic beauty.

11.1. Modular invariants

Let $\mathcal{L}$ be a lattice in the complex plane, primarily an aggregate of points Ω, written here in terms of a basis as

$$(11.1.1) \quad \mathcal{L} = \{\Omega\} = [\omega_1, \omega_2], \qquad \Omega = n_1\omega_1 + n_2\omega_2 \quad (n_i \in \mathbf{Z}).$$

We require only that ω_1/ω_2 be nonreal. (In studying quadratic forms in Section 8.1, we required that Im $\omega_2/\omega_1 > 0$ for orientation, but we do not yet require that here. In fact, we do not yet require that ω_1 or $\omega_1/$

ω_2 be algebraic.) We denote the action of a matrix $A \in G^*$ on $\mathcal{L}$ (through a given basis, of course) by

$$(11.1.1a) \qquad A[\omega_1, \omega_2] = [a\omega_1 + b\omega_2, c\omega_1 + d\omega_2]$$
$$A = \begin{pmatrix} a & b \\ c & d \end{pmatrix}, \qquad n = \det A = ad - bc > 0.$$

11.1.2. Definition. *Let $f(\tau)$ be a modular form of degree m, with $m \geq 0$ (where $m = 0$ is interpreted as describing a modular function). Thus,*

$$(11.1.2a) \qquad (t\tau + u)^{-2m} f(M\tau) = f(\tau), \qquad M \in \Gamma$$
$$M = \begin{pmatrix} r & s \\ t & u \end{pmatrix}, \qquad ru - st = 1.$$

If ω_1 and ω_2 are two complex numbers with a nonreal ratio, we define the "homogeneous form" for the function $f(\omega)$ as

$$(11.1.2b) \qquad f(\omega_1, \omega_2) = f(\omega_1/\omega_2)\omega_2^{-2m} \quad if \, Im \, \omega_1/\omega_2 > 0$$

whereas the condition $f(\omega_1, \omega_2) = f(\omega_2, \omega_1)$ extends the definition otherwise. [This is the converse of the derivation of $G_{2m}(\tau)$ from $G_{2m}(\omega_1, \omega_2)$ in Section 10.1.] As a result, for $M \in \Gamma$,

$$(11.1.2c) \qquad f(M[\omega_1, \omega_2]) = f(\omega_1, \omega_2).$$

Analogously with Section 10.4, we define for $f_A(\tau)$,

$$(11.1.2d) \quad f_A(\omega_1, \omega_2) = n^{2m} f(A[\omega_1, \omega_2])/f(\omega_1, \omega_2) \qquad det \, A = n > 1.$$

11.1.3. Theorem. *If $\mathcal{L}$ is a lattice and $f(\tau)$ is a modular form of dimension $m \, (\geq 0)$, then for any basis we define*

$$f(\mathcal{L}) = f(\omega_1, \omega_2).$$

In this definition f depends only on the lattice $\mathcal{L}$ and not on the choice of basis. (It is therefore called a "modular invariant.")

11.1.4. Definition. *Two lattices $\mathcal{L}$, $\mathcal{L}'$ (or modules) are "similar" when*

$$\mathcal{L} = \alpha \mathcal{L}' \qquad (\alpha \in C^*)$$

(by a biunique correspondence of elements of each). All similar lattices form an equivalence class c.

11.1.5. Theorem. *The modular invariant $j(\mathcal{L})$ biuniquely maps the equivalence classes of lattices under similarity onto the finite complex plane. [Thus, $j(c)$ will be used for the common value of a class c of $\mathcal{L}$.]*

11.1.6. Definition. *The "endomorphisms" $R(\mathcal{L})$ of a lattice are the set of complex numbers $\{\alpha\}$ such that $\alpha = 0$ or $\alpha\mathcal{L}$ is a sublattice of $\mathcal{L}$. These endomorphisms form a "ring" under the natural operations on $\mathcal{L}$, $\alpha\beta\mathcal{L} = \alpha(\beta\mathcal{L})$, $(\alpha + \beta)\mathcal{L} = \alpha\mathcal{L} + \beta\mathcal{L}$ (elementwise set addition). Thus, $R \supseteq \mathbf{Z}$. The same definition gives $R(c)$, the endomorphisms of a class of lattices. We say that "complex multiplication" exists if the ring R is strictly larger than $\mathbf{Z}$.*

11.1.7. Theorem. *In the case of complex multiplication, the elements of R that are not in $\mathbf{Z}$ are nonreal. Indeed, $R = \mathbf{Z}[\xi]$, where ξ is a quadratic integer. Thus,*

$$(11.1.7a) \qquad\qquad R = O_f = [1, f\omega]$$

where $O = [1, \omega]$ is the integral domain of some $k = \mathbf{Q}(\sqrt{d})$, where $d < 0$. The lattice $\mathcal{L}$ is proportional to some module in k (or 0).

Proof. If $\alpha\mathcal{L}$ is a sublattice of $\mathcal{L}$ (of, say, index n),

$$\alpha\omega_1 = a\omega_1 + b\omega_2$$
$$ad - bc = n$$
$$\alpha\omega_2 = c\omega_1 + d\omega_2$$

$$\det\begin{pmatrix} \alpha - a & -b \\ -c & \alpha - d \end{pmatrix} = 0.$$

Clearly, α is an algebraic integer [and is complex because $\omega_2/\omega_1 = (\alpha - a)/b$, which cannot be real]. $\square$

11.1.8. Corollary. *Every (nonsingular) lattice $\mathcal{L} \subset k$ has a complex multiplication ring in k.*

Proof. If $\mathcal{L} = [\omega_1, \omega_2]$ and $\omega_1/\omega_2 = \tau$ satisfies an equation of denominator F, that is, $F\tau^2 + G\tau + H = 0$, then

$$F\tau[1, \tau] \subseteq [1, \tau], \quad \text{hence } \mathfrak{F}\tau\mathcal{L} \subset \mathcal{L}. \square$$

11.1.9. Corollary. *Every lattice $\mathcal{L} \subset k$ is proportional to a regular ideal $\mathfrak{m}$ in the multiplier ring $R(\mathcal{L})$, that is, an ideal of norm prime to f. Thus, if $\alpha \in \mathbf{Q}(\sqrt{d})$, then $j(\alpha) = j(\mathfrak{m})$ for some such ideal (see Lemma 8.1.10).*

11.1.10. Lemma. *Let $\mathfrak{m}$ and $\mathfrak{q}$ be ideals in O_f that are regular (i.e., prime to f) and let $\mathfrak{q}$ be primitive (i.e., $\mathfrak{q}/t$ is not in O_f if $t > 1$ and $t\epsilon\mathbf{Z}$). Let $N\mathfrak{q}$*

= q. *Then in terms of some basis* $m = [\omega_1, \omega_2]$, $mq = Sm$ *with S a primitive matrix of norm q. In particular, S can always be chosen in canonical form [see equation (10.41b)].*

Proof. Since $m \supset qm$, the ideal qm is a multiple but also a sublattice of m described by a matrix $S[\omega_1, \omega_2] = qm$, $\det S = q = [m : qm]$. But $S[\ldots]$ can be replaced by $ES[\ldots]$ for $E \in \Gamma$ without affecting the ideal qm (just by changing its basis). Thus, ES can be chosen as canonical [see equation (10.2.6a)]. □

11.1.11. Theorem. *The modular invariant $j(\mathcal{L})$ for $\mathcal{L} \subset k$ is an algebraic integer.*

Proof. Let λ be an element of $R(\mathcal{L})$ of nonsquare norm n [i.e., $\lambda = f\sqrt{d}$ if $d \neq -4$ or $f(1 + i)$ if $d = -4$]. Then $\lambda[\omega_1, \omega_2] = A[\omega_1, \omega_2]$ for $\det A = n$. Then $j(\mathcal{L})$ is a root of $J_n(j, j) = 0$. We now need to see that this equation is monic. This follows from the q-expansion of

$$J_n(j, j) = \prod_{\psi(n)} (j(\tau) - j(a\tau + b)/d)).$$

If $a \neq d$ (as follows when n is nonsquare), then each factor is like a power of q times a root of unity. So the product is monic in powers of q. □

11.1.12. Definition. *Let $c_1, \ldots, c_h$ be the (regular) ideal classes of the ring O_f in $k = \mathbf{Q}(\sqrt{d})$, where $d = d_1 f^2$ (with c_1 the unit class). Then*

(11.1.12a) $$j(c_1), \ldots, j(c_h)$$

are called the "class invariants" of O_f [$h = h(d_1 f^2)$]. The "class polynomial" (or equation) refers to

(11.1.12b) $$H_d(t) = (t - j(c_1)) \cdots (t - j(c_h))$$

clearly lying in $\mathbf{C}[t]$. (Ultimately it will be shown to lie in $\mathbf{Q}[t]$ and hence $\mathbf{Z}[t]$, making it the defining polynomial for each invariant.)

11.1.13. Theorem. *The class invariants comprise all the values of $j(\alpha)$ for α equal to an imaginary quadratic irrationality.*

11.1.14. Weber's theorem. *The class invariants are the roots of an equation of degree h over k. They each define the same field K/k of degree h*

that is also relatively abelian with the Galois group isomorphic to the ideal class group. Finally, K is the unit ring class field over k for the ring O_f; that is, p splits completely from $\mathbf{Q}$ to K if and only if p splits into $\mathfrak{p}$ and $\mathfrak{p}'$ from $\mathbf{Q}$ to k and each of these factors in k is an ideal in the unit class c_1.

The proof occupies this section and the next. As a first step, we construct the Galois closure

$$(11.1.15a) \qquad K = \bar{k}(j(c_1), \ldots, j(c_h)) \qquad [k = \mathbf{Q}(\sqrt{d})].$$

[Of course, we hope to show that $K = k(j(c_m))$ for each m, at the very least.] We can establish class field properties with the exclusion of any finite set of primes. We systematically exclude prime divisors of

$$(11.1.15b) \qquad D = 2fd \prod_m d(j(c_m)), \qquad 1 \leq m \leq h,$$

where $d(\theta)$ denotes the root discriminant, as in Section 4.1. For the rest of the proof of Weber's theorem, it shall be assumed that $p \nmid D$.

11.1.16. Lemma. *Let $p = \mathfrak{p}\mathfrak{p}'$ in k and let $\mathfrak{P}$ (in K) divide $\mathfrak{p}$. Then*

$$j(\mathfrak{p}c) \equiv j(c)^p \quad or \quad j(\mathfrak{p}c)^p \equiv j(c) \bmod \mathfrak{P}$$

for an ideal class c.

Proof. This is an immediate consequence of Lemma 11.1.10 and Theorem 10.4.12. □

11.1.17. Lemma. *Once it is known that K/k is abelian, we can conclude the part of Theorem 11.1.14 concerning the class field property.*

Proof. Take $(K/k/\mathfrak{p}) = S$ where $\mathfrak{P}$ (in K) | $\mathfrak{p}$ (in k) | p, all prime ideals of degree 1 for convenience. Then by the Artin symbol,

$$A^S \equiv A^p \bmod \mathfrak{p} \qquad (A \in O_K).$$

Now conjugates of A are determined by conjugation on $j(c_m)$. So we can restrict A to $j(c_m)$. It follows that $\{S = 1\} \Leftrightarrow \{\mathfrak{p}$ is principal$\}$. [We must exclude the various p dividing D in equation (11.1.15b) so that, for instance, the values $j(c_m)$ can each be identified by their residues mod p.] □

Despite the appearance of Lemma 11.1.16, there is no option,

$$(11.1.18) \qquad\qquad j(\mathfrak{p}'c) \equiv j(c) \bmod \mathfrak{P}$$

(the second possibility in the lemma, if we replace c by $c\mathfrak{p}'$). To prove this stronger result is, surprisingly, much harder and it will be reserved to the next section. (The congruence is also valid mod $\mathfrak{p}$ because it is valid for all $\mathfrak{P}$ that divide $\mathfrak{p}$.)

11.1.19. Lemma. *If the relation (11.1.18) is shown, then it follows that $H_d(t)$ lies in $\mathbf{Q}[t]$ and K/k is abelian (and Weber's theorem follows).*

Proof. Consider the field K/k, for K the normal field over k generated by the roots of the class polynomial. By Chebotarev's theorem (7.5.7) it follows that all elements S of Gal K/k are generated as Frobenius automorphisms of infinitely many primes $\mathfrak{P}$ in K of degree 1; that is,

$$A^S \equiv A^p \bmod \mathfrak{P} \qquad (A \in O_k).$$

Applying this to the h class invariants $j(c)$ (which are distinct modulo $\mathfrak{P}$), we have

$$j(c)^S = j(c\mathfrak{p}').$$

This states a remarkable fact; Gal K/k is isomorphic to the class group if S corresponds to the class of $\mathfrak{p}'$ (uniquely determined by $\mathfrak{P}$, as $N_{K/k}\mathfrak{P}$ $= \mathfrak{p}$ and $\mathfrak{p}' = p/\mathfrak{p}$). Then K/k is easily seen to be of degree h and abelian. $\qquad\qquad\square$

11.1.20. Corollary. *From the preceding proof, $j(c_i)$ {i.e., $j(f\omega)$, where O_f $= [1, f\omega]$} generates over $\mathbf{Q}$ the maximal real subfield of k. It is generally nonabelian over $\mathbf{Q}$ (see Exercise 11.1).*

11.1.21a. Illustration: class number unity. There are 13 rings of discriminant d (not necessarily fundamental) for which $h(d) = 1$. These are exactly the cases where $j(c_i)$ is rational by the preceding corollary. We write $j\{d\}$ as this value for convenience. The formula (10.1.18) is sufficiently accurate to check the cases (beyond the first two, which follow from the definition of j):

$$j\{-3\} = 0, \quad j\{-4\} = 12^3, \quad j\{-7\} = -15^3, \quad j\{-8\} = 20^3,$$
$$j\{-11\} = -32^3, \quad j\{-12\} = 2 \cdot 30^3, \quad j\{-16\} = 66^3,$$
$$j\{-19\} = -96^3, \quad j\{-27\} = -3 \cdot 160^3, \quad j\{-28\} = 255^3,$$
$$j\{-43\} = -960^3, \quad j\{-67\} = -5280^3, \quad j\{-163\} = -640{,}320^3.$$

Thus, $H_{-3}(t) = t$, $H_{-4}(t) = t - 12^3$, $H_{-7}(t) = t + 15^3$, and so on.

11.1.21b. Illustration: higher class numbers. Although $j(c)$ is not rational in general, the symmetric functions are always both rational and integral. Hence, they too can be calculated numerically: For $d = -20$, $h = 2$, so $j(c_1) = j(\sqrt{-5})$ and $j(c_2) = j((1 + \sqrt{-5})/2)$;

$$j(\sqrt{-5})j((1 + \sqrt{-5})/2) = -880^3$$
$$j(\sqrt{-5}) + j((1 + \sqrt{-5})/2) = 158{,}000$$

thus, $j(\sqrt{-5}) = (50 + 26\sqrt{5})^3$, $j((1 + \sqrt{-5})/2) = (50 - 26\sqrt{5})^3$, and $H_{-20}(t) = (t^2 - 2 \cdot 79 \cdot 10^3 t - 880^3)$.

Likewise, for $d = -24$, $h = 2$, similar numerical approximations reveal that $j(\sqrt{-6}) = 12^3(1 + \sqrt{2})^2(5 + 2\sqrt{2})^3$, whereas the other class, $j(\sqrt{-6}/2)$, gives the conjugate value, as usual. Thus,

$$H_{-24}(t) = (t^2 - 12^3 \cdot 2798t + 12^6 \cdot 17^3).$$

For $d = -15$, $h = 2$, $j((1 + \sqrt{-15})/2)$ and $j((1 + \sqrt{-15})/4)$ are the conjugates $135\sqrt{5}((3 \pm \sqrt{5})/2)(4 \pm \sqrt{5})^3$. Thus,

$$H_{-15}(t) = t^2 - 3^3 \cdot 5^2 \cdot 283t - 3^6 \cdot 5^3 \cdot 11^3.$$

A more difficult example is for $d = -23$, $h = 3$. There

$$j((1 + \sqrt{-23})/2), \quad j((1 + \sqrt{-23})/4), \quad j((1 - \sqrt{-23})/4)$$

are the three conjugates of $-125\xi^{12}(2\xi - 1)^3(3\xi + 2)^3$, where $\xi^3 - \xi - 1 = 0$. [The roots must correspond to the $j(c)$ values in such a way that the first is real and the others are complex conjugates.] A rather lengthy calculation is involved here but we give only the final polynomial:

$$H_{-23}(t) = t^3 + 250 \cdot 13{,}967t^2 + 5^6 \cdot 329{,}841t + 5^9 \cdot 187^3.$$

The preceding examples account for all H_d for $|d| < 31$.

11.1.22. Remark. We have yet to prove equation (11.1.18) and the rationality of the class polynomial. So far, we do know (by Lemma 11.1.19) that K contains the desired unit ring class field over k.

11.2. Weber's theorem

First we recall that the class invariants $j(c)$ (also called "singular moduli") are roots of the class polynomials $H_d(t)$. We now show that these polynomials are factors of $J_n(t, t)$.

We note that for $\mathfrak{m}$ a regular ideal in O_f [$d = d_1 f^2$, $k = \mathbf{Q}(\sqrt{d})$, $d < 0$ as usual] with S a primitive matrix ($\det S = n \geq 1$), the resulting

$S\mathfrak{m}$ is a submodule of $\mathfrak{m}$. Yet there is no easy algorithm for verifying in advance that $S\mathfrak{m}$ has the more special form $\lambda\mathfrak{m}$ or $\mathfrak{q}\mathfrak{m}$ for an integer λ or an ideal $\mathfrak{q}$ in O_f. An obvious necessary condition is that n actually equals a norm of λ or of $\mathfrak{q}$; beyond this, very little can be said.

11.2.1. Lemma. *Let $\mathfrak{m}$ be a primitive regular ideal in O_f with given basis and let λ be a primitive integer of O_f of norm n (≥ 1), but not necessarily prime to f. Then $j(\mathfrak{m}) = j(A\mathfrak{m})$ for some primitive matrix of determinant n in a unique canonical form $\begin{pmatrix} a & b \\ 0 & d \end{pmatrix}$.*

Proof. We have $\lambda\mathfrak{m} = S\mathfrak{m}$ for $\det S = n$, so

$$j(\mathfrak{m}) = j(\lambda\mathfrak{m}) = j(S\mathfrak{m}) = j(MS\mathfrak{m}), \qquad M \in \Gamma.$$

The correct choice of M produces the canonical $MS = A$. $\qquad\square$

11.2.2. Lemma. *If $j(\mathfrak{m}) = j(S\mathfrak{m})$ for $\det S = n$, S primitive, then a primitive λ exists such that $\lambda\mathfrak{m} = S\mathfrak{m}$ and $N\lambda = n$.*

Proof. Here λ is an eigenvalue of S. $\qquad\square$

11.2.3. Lemma. *The canonical matrix for an integer λ for Lemma 11.2.1 is biuniquely determined by the ideal (λ) in O_f. (All λ that lead to the same A are associates in O_f.)*

Proof. If $\lambda_i\mathfrak{m} = S_i\mathfrak{m}$ $(N\lambda_i = \det S_i = n)$ and $M_iS_i = A_i$ for $M_i \in \Gamma$, then $A_1A_2^{-1} = M_1S_1S_2^{-1}M_2^{-1}$, which is an element of Γ if and only if $S_1S_2^{-1} \in \Gamma$. But this is precisely when $\lambda_1\mathfrak{m} = \lambda_2\mathfrak{m}$, or equivalently, λ_1/λ_2 is a unit of O_f [the complex multiplication ring $R(\mathfrak{m})$]. $\qquad\square$

11.2.4. Theorem. *We can factor $J_n(t, t) = u \prod H_d(t)^{e(d)}$, $n > 1$, where the product is taken over all those discriminants d of rings O_f $(d = d_1 f^2)$ in which $n = N\lambda$ is solvable for λ primitive in O_f. (We do not require that n be prime to f.) Here $e(d)$ is the number of nonassociated λ [or ideals (λ)] in O_f and $u \in \mathbf{Z}$ $(u = \pm 1$ if n is not a perfect square).*

Proof. This follows from the fact that $J_n(t, t)$ has roots where $j(\mathfrak{m}) = j(A\mathfrak{m})$ for primitive A of determinant n; and for each d the roots are counted by the number of canonical A, namely $e(d)$. $\qquad\square$

11.2.5. Illustrations. As hard as it is to compute $H_d(t)$, it is much harder to compute $J_n(t, j)$. Very few results are known. Even the simplest is a major computation;

(11.2.5a)
$$\begin{aligned}
J_2(t, j) = {}&-(t^2 - {}^5j)(j^2 - t) + 2^4 \cdot 93jt(j + t) \\
&-2^4 \cdot 3^4 \cdot 5^3(j^2 + t^2) + 2^8 \cdot 7 \cdot 61 \cdot 737jt \\
&+2^8 \cdot 3^7 \cdot 5^6(j + t) - 2^{12} \cdot 3^9 \cdot 5^9 = 0.
\end{aligned}$$

It is now easy to verify from Illustration 11.1.21a that

(11.2.5b)
$$\begin{aligned}
J_2(t, t) &= -(t - 12^3)(t - 20^3)(t + 15^3)^2 \\
&= -H_{-4}H_{-8}H_{-7}^2.
\end{aligned}$$

These factors come from the norm representations

$$2 = N(1 + i) = N(\sqrt{2}) = N(1 \pm \sqrt{-7})/2).$$

(Here $1 - i$ does not count additionally because it is an associate of $1 + i$.)

We can represent such factorizations symbolically by the degrees:

(11.2.5c) $\deg J_n(t, t) = J[n], \quad \deg H_d(t) = h(d).$

The first few cases are now listed.

(11.2.5d)
$$\left\{\begin{aligned}
J[2] &= h(-4) + h(-8) + 2h(-7) = 4 \\
J[3] &= h(-3) + h(-12) + 2h(-8) + 2h(-11) = 6 \\
J[4] &= h(-16) + 2h(-7) + 2h(-12) + 2h(-15) = 9 \\
J[5] &= h(-20) + 2h(-4) + 2h(-11) + 2h(-16) \\
&\quad + 2h(-19) = 10 \\
J[6] &= h(-24) + 2h(-8) + 2h(-15) + 2h(-20) \\
&\quad + 2h(-23) = 18 \\
J[7] &= h(-7) + h(-28) + 2h(-3) + 2h(-12) \\
&\quad + 2h(-19) + 2h(-24) + 2h(-27) = 14 \\
J[8] &= h(-32) + 2h(-7) + 2h(-16) + 2h(-23) \\
&\quad + 2h(-28) + 2h(-31) = 20.
\end{aligned}\right.$$

Some details are in order: For $J[7]$, $7 = N(2 + \sqrt{-3}) = N((1 + 3\sqrt{-3})/2) = N((5 + \sqrt{-3})/2)$; all are associates when $d = -3$. When $d = -12$, only the first is valid, and when $d = -27$, only the second. (The conjugates, of course, are nonassociates.) Likewise, for $J[8]$, $8 = N(2 + 2i)$, but $2 + 2i$ is not primitive when $d = -4$, only when $d = -16$. But then, $2 - 2i$ is not associated, since i is no longer an integer of the ring. The new class numbers not mentioned in Section 11.1 are $h(-31) = 3$ and $h(-32) = 2$. (Note that new class numbers may be computed in this fashion.)

11.2.6. Theorem. *The class polynomial $H_d(t)$ is the gcd (in the ring $\mathbf{Q}[t]$ or $\mathbf{Z}[t]$) of those $J_n = J_n(t, t)$ for which $n = N\lambda$ for λ primitive in O_f. Thus, $H_d(t)$ lies in $\mathbf{Z}[t]$.*

Proof. We exclude the small cases, so we can assume $|d| > 4$. Thus,

$$H_{-3} = \gcd(J_3, J_7), \qquad H_{-4} = \gcd(J_2, J_5).$$

There are two cases otherwise (with $d = d_1 f^2$), namely:

1. For $d_1 \equiv 1 \bmod 4$, $O_f = [1, f\omega]$, $\omega = (1 + \sqrt{d_1})/2$.
2. For $d_1 \not\equiv 1 \bmod 4$, $O_f = [1, f\omega]$, $\omega = (1 + \sqrt{d_1}/2)$.

In either case, set $d' = N(f\omega)$. Then in these cases it can be seen that

1. $H_d = \gcd(J_{|d|}, J_{d'})$ f odd
   $H_d = \gcd(J_{|d/4|}, J_{d'})$ f even.
2. $H_d = \gcd(J_{|d/4|}, J_{d'})$.   □

From the properties of $\Delta_S(\tau)$ in Section 10.4, for an imaginary surd α, $\Delta_S(\alpha)$ is an integer in the number field $\mathbf{Q}(j(\alpha), j(S\alpha))$. Also for the case of determinant p (prime), if S runs over the canonical matrices A, then by Corollary 10.4.11,

(11.2.7) $$\Pi \, \Delta_A(\alpha) = p^{12}.$$

The same holds for $\Delta_S(\mathfrak{m})$ for $\mathfrak{m}$ an ideal, under the homogeneous definition (11.1.2).

11.2.8. Theorem. *Let $\mathfrak{m}$ and $\mathfrak{p}$ be regular ideals in O_f with given basis for $\mathfrak{m}$, and $N\mathfrak{p} = p$ (prime). Then in the normal field K of $j(\mathfrak{m})$ and $\Delta_S(\mathfrak{m})$, if S is of determinant p, these ideal factorizations are valid:*

$$\Delta_S(\mathfrak{m}) = \begin{cases} \mathfrak{p}^{12}, & S\mathfrak{m} = \mathfrak{p}'\mathfrak{m} \\ \mathfrak{p}'^{12}, & S\mathfrak{m} = \mathfrak{p}\mathfrak{m} \\ (1), & otherwise. \end{cases}$$

Proof. Let $\mathfrak{g}$ be prime to $\mathfrak{p}$ ($N\mathfrak{g} = g$), satisfying $\mathfrak{g}\mathfrak{p} = (\gamma)$ (for ideals regular in O_f), and $\gamma\gamma' = N\gamma = pg$. Let $\mathfrak{p}\mathfrak{m} = A\mathfrak{m}$, $\mathfrak{g}\mathfrak{p}\mathfrak{m} = BA\mathfrak{m}$, for matrices A, B. By definition,

$$\Delta_{BA}(\mathfrak{m}) = g^{12}p^{12}\Delta(BA\mathfrak{m})/\Delta(\mathfrak{m}) = \gamma^{-12}g^{12}p^{12}$$

since $BA\mathfrak{m} = \mathfrak{p}\mathfrak{g}\mathfrak{m} = \gamma\mathfrak{m}$ and $\Delta(\gamma\mathfrak{m}) = \gamma^{-12}\Delta(\mathfrak{m})$. Therefore, with $\gamma\gamma' = gp$, $\gamma'^{12} = \mathfrak{g}'^{12}\mathfrak{p}'^{12} = \Delta_{BA}(\mathfrak{m}) = \Delta_B(A\mathfrak{m})\,\Delta_A(\mathfrak{m})$. The theorem follows from the product formula (11.2.7), since Δ_B must be prime to p and Δ_A must be prime to q.   □

It was shown in Section 10.4 that in terms of function fields, $\Delta_A(\tau)$ lies in $\mathscr{F}_A = \mathbf{Q}(j(\tau), j_A(\tau))$ (and likewise for the number fields that arise from singular moduli). But we can write

$$j_A(\tau) \in \mathbf{Q}(j(\tau), \Delta_A(\tau)).$$

This is true because δ_A is primitive; that is, the conjugates for each canonical A are different. What we have here is, however, a more complicated situation, since we want to substitute algebraic numbers for t and this may cause primitivity to fail. In other words, $d(\Delta_A) \neq 0$ in the function field, but this might not be true in the number field. Similarly, by equations (10.4.9d), note that if $A = \begin{pmatrix} p & 0 \\ 0 & 1 \end{pmatrix}$, then $\Delta_A(\tau)$ has a q-series divisible by p, but it may become a unit by the preceding theorem if a "general" $\mathfrak{m}$ is substituted for t. (See Theorem 11.2.8.) What we are really doing is distinguishing between congruences "mod. p" valid for a q-series and "mod p" valid for numerical values.

11.2.9. Lemma. *For p prime, in Theorem 10.4.13, when we substitute an imaginary quadratic surd for τ, the relation*

$$\Phi_p(t, j) \equiv t^{p+1} + g(j)t \bmod p$$

yields a numerical value of $g(j)$ relatively prime to p.

Proof. Here $g(j)$ is a symmetric function over the canonical A, $g(j) = \Pi \Delta_A(\Sigma\, 1/\Delta_A)$. Hence, by Theorem 11.2.8, $g(j)$ has (ideal) summands divisible by (p) except for two terms equal to $\mathfrak{p}^{12}$ and $\mathfrak{p}'^{12}$ (of course $\mathfrak{p} \neq \mathfrak{p}'$). $\qquad\square$

11.2.10. Theorem. *Let $f_A(\tau) \in \mathbf{Z}[j(\tau), j_A(\tau)]$ and suppose that $f_A(\tau)$ has a q-series divisible by p when $A = \begin{pmatrix} p & 0 \\ 0 & 1 \end{pmatrix}$; that is, $A\tau = p\tau$. Then for each of the $[\psi(p) =]\, p + 1$ canonical A and fixed $a_m(j) \in \mathbf{Z}[j]$,*

$$(11.2.10a) \qquad f_A(\tau)\Phi_p'(\Delta_A(\tau), j(\tau)) = \sum_{m=0}^{p} a_m(j)\, \Delta_A(\tau)^m$$

[where $\Phi_p'(t, j) = \partial\Phi/\partial t$]. Furthermore, $a_o(j)$ has coefficients divisible by p.

Proof. Proceeding directly from Theorem 10.4.14, we observe that $\Delta_A(\tau)$ also has a q-series divisible by p when $A\tau = p\tau$. $\qquad\square$

Weber's theorem finally reduces to the choice of the "correct" Frobenius automorphism [see equation (11.1.18)].

11.2.11. Theorem. *Let* m *be a regular ideal and* p *a prime ideal of norm* p *in* O_f *divisible by* $\mathfrak{P}$ *in* K, *then*

$$j(p'm) \equiv j(m)^p \ mod \ \mathfrak{P}.$$

Proof. Apply Theorem 11.2.10 to $f_A = j_A(\tau) - j(\tau)^p$. Then we use the special A (of determinant p) for which

(11.2.11a) $$Am = p'm, \qquad \Delta_A(m) \equiv 0 \ mod \ p$$

by Theorem 11.2.8. Also, by Lemma 11.2.9, for this A,

$$\Phi_p'(\Delta_A(\tau), j(\tau)) \equiv \Delta_A(\tau)^p + g(j) \not\equiv 0 \ mod \ p.$$

Therefore $f_A(m) \equiv 0 \ mod \ \mathfrak{P}$. (The result also holds more easily in the case of $p = p'$.)

11.2.12. Remark on Hilbert class fields. We note that Theorem 11.2.8 provides us with an explicit Hilbert class field. We have a field where each p^{12} is principal. Therefore, for any ideal class c, c^{12} is also principal. Thus we have a Hilbert class field if h is prime to 6. There is a more useful theorem based on a result of Fricke that $\Delta_A^{1/24}$ lies in the same field as Δ_A if det A is a square prime to 6. We do not prove this here, but we can state the result:

11.2.12a. Theorem (Fricke). *If* m *and* b *are regular ideals of* O_f *where a basis is assigned to* m *and* b *is prime to 6, then if B is the matrix for which* $Bm = mb'^2$, *it follows that* $\Delta_B(m)$ *generates the same ideal as* b *in* $Q(j(m), j_B(m))$.

11.3. Iteration and the icosahedron

We recall that in solving an equation (say, over Q) by Galois theory we must find all its roots at once (i.e., we must construct the splitting field or the resolvent equation). We actually express all roots in terms of a generator of the normal field. The equation is called "solvable" when the generator is expressible in terms of radicals. Within the normal field, the roots of the resolvent equation must be rational functions of one another. In his classic work, Klein showed that the roots of the resolvent can be expressed in terms of one another by groups of linear fractional transformations that denote the rotations of regular solids. Of these, the icosahedron (used for the quintic equation) became the most conspicuous and symbolic. One historical fact not well appreciated today is that this work was motivated by the process of expressing relations between the roots of the modular equation.

Consider, for instance, the modular equation $J_b(j(b\tau), j(\tau))$ (see Section 11.1). Here, given $j(\tau)$, the roots are a set of M conjugates to $j(b\tau)$, namely,

(11.3.1a) $j_r(\tau) = j((A\tau + B)/C, \qquad r = 1, \ldots, M$

(11.3.1b) $AC = b, \qquad 0 \le B < C, \qquad 0 < A, \qquad (A, B, C) = 1$

(11.3.1c) $M = b \, \Pi \, (1 + 1/p)$

(where p runs over the prime divisors of b). The roots are interconnected by the rotations of a regular euclidean solid for $b = 2, 3, 4, 5$ (and by more complicated groups for other b), but the roots must first be expressed in terms of suitable parameters. Aside from the obvious geometric interest, this process is important because it expresses iteration of the type

(11.3.1d) $j(\tau), j(b\tau), j(b^2\tau), \ldots, j(b^t\tau), \ldots$

and thus this process leads to the construction of ring class fields of discriminant $-\Delta b^{2t}$ as t varies.

This iteration procedure can be illustrated by an example. Klein showed that $j(2\tau)$ can be calculated from $j(\tau)$ by solving

(11.3.2a) $j(\tau) = 64(\eta + 3)^3/(\eta - 1)^2,$
                   $j(2\tau) = 64(\eta_0 + 3)^3/(\eta_0 - 1)^2$

(11.3.2b) $(\eta - 1)(\eta_0 - 1) = 1$

[i.e., by eliminating η and η_0 from equations (11.3.2a–b)]. The problem is that we are simultaneously calculating $j(\tau/2)$ and $j((\tau + 1)/2)$ with $j(2\tau)$. In a numerical calculation there would be no difficulty distinguishing these roots. [e.g., $j(2\tau)$ might be the largest]. Abstractly, however, we would have a problem: For instance, the system (11.3.2a, b) leads to some algorithm for writing

(11.3.2c) $j(\tau) \to \eta \to \eta_0 \to j(2\tau)$

provided we agree on which root to take in the first step, $j(\tau) \to \eta$. (We defer a discussion of these formulas until later.)

This process leads to a paradoxical situation, in that whatever algorithm is used in $j(\tau) \to \eta$ must *not* be reversible if we want to iterate as in (11.3.1d). If the process of (11.3.2c) were reversible, we would find that starting with $j(2\tau)$, we obtain $j(\tau)$ and not $j(4\tau)$. [Note that $j(\tau)$ and $j(4\tau)$ are conjugates if we start with $j(2\tau)$.] To progress from $j(\tau) \to j(2\tau) \to j(4\tau)$, we must destroy reversibility. It is done by inserting a new step in the chain (11.3.2c), in which the root η is replaced by η', a conjugate. Thus,

$$(11.3.2d) \qquad j(\tau) \to \eta \to \eta' \to \eta_0 \to j(2\tau).$$

The system (11.3.2a, b) becomes the following:

$$(11.3.3a) \qquad j(\tau) = 64(\eta + 3)^3/(\eta - 1)^2$$
$$j(2\tau) = 64(\eta_0 + 3)^3/(\eta_0 - 1)^2$$
$$(11.3.3b) \qquad (\eta \to \eta'), (\eta' - 1)(\eta_0 - 1) = 1.$$

The transition $(\eta \to \eta')$ is now the vital step, which we state (leaving the proofs for later). We introduce new variables z, z':

$$(11.3.4a) \qquad \eta = z^2, \qquad \eta' = z'^2$$

and it is the variables z and z' that are related by a "rotation" on the z-sphere,

$$(11.3.4b) \quad z' = (z + 3)/(z - 1) \quad \text{or} \quad (z - 1)(z' - 1) = 4.$$

If we now take equations (11.3.3a, b) and (11.3.4a, b) and eliminate five variables η, η_0, η', z, and z', we have an irreversible procedure that produces the iteration $j(\tau) \to j(2\tau) \to j(4\tau) \to \dots$.

Actually, the iteration can be performed in the variables η and z [without returning to $j(\tau)$ each time]. Indeed, the equations (11.3.3b) and (11.3.4a, b) lead to the procedure

$$(11.3.5a) \quad j(2^t\tau) = 64(\eta^{(t)} + 3)^3/(\eta^{(t)} - 1)^2$$
$$(11.3.5b) \quad \eta^{(t+1)} = F_2(\eta^{(t)})$$
$$(11.3.5c) \quad F_2(\eta) = [(5\eta - 9) + (\eta + 3)\eta^{1/2}]/8(\eta - 1).$$

Equivalently, we can define

$$(11.3.6a) \qquad z^{(t)} = \sqrt{\eta^{(t)}}$$
$$(11.3.6b) \qquad z^{(t+1)} = \sqrt{G_2(z^{(t)})}$$
$$(11.3.6c) \qquad G_2(z) = (z + 3)^2/8(z + 1).$$

This is the formula (1.10) used in Chapter 1 for testing whether or not $j(2^t i)$ is represented by a rational modulo p.

At this point we call attention to Tables 1 through 6, which summarize the iteration process for $b = 2$, 3, 4, and 5. The tables are almost self-explanatory, but the following comments may help: First, the special values of b for which Klein originally expressed $j(\tau)$ and $j(b\tau)$ parametrically were chosen from those for which the modular equation $J_b(z, w) = 0$ defines a Riemann surface of genus 0, namely,

$$b = 2, 3, 4, 5, 6, 7, 8, 9, 10, 12, 13, 16, 18, 25.$$

Because we are working with resolvents, however, we make the further restriction that the normal field defined by $j(b\tau)$ over $j(\tau)$ is also of genus 0. These values are $b = 2, 3, 4, 5$ finally. For each of these cases, we introduce a variable η upon which the iteration is to be performed. [This way, we do not return to $j(\tau)$ each time, as in equations (11.3.5a, b, c).] By definition of M (the degree), the parametrization of $j(\tau)$ and $j(b\tau)$ must be of degree M in n. We introduce a new variable z:

(11.3.7a) $z^b = \eta$ $(b = 2, 3, 4)$.

Then (as we later verify) it is upon the z-sphere that the actions of G, the euclidean rotation group, take place. The group is of order

(11.3.7b) $|G| = bM$ $(b = 2, 3, 4)$.

When $b = 5$, however, it is necessary to introduce a more complicated variable

(11.3.8a) $z^5 - 1/z^5 = -11 + 125/\eta$

so that now the group G that acts on the z-sphere satisfies

(11.3.8b) $|G| = 2bM \, (= 60, \, b = 5)$.

The transformations are shown in Tables 1, 2, 3, and 4. The coefficients lie in $\mathbf{Q}(\omega)$, where

(11.3.9) $\omega = \exp 2\pi i/b$ $(b = 2, 3, 4, 5)$.

The groups G are familiar (see Table 1).

The case $b = 5$ is a matter of more special concern. By contrast, if $b = 3$, any primitive form of discriminant $-\Delta b^{2t}$ has the genus property that if the odd prime p is represented by this form, then $p \equiv 1$ mod b. In effect, this means that ω is represented by an integer modulo p. This is not true when $b = 5$; there the best we can conclude is that $p \equiv \pm 1$ mod 5. (Unfortunately, $p \equiv 1$ is required to represent ω by an integer modulo p.) To keep the computation in strictly rational terms, we introduce an alternative parameter to z (just for $b = 5$), namely λ (which is always representable by an integer modulo p),

(11.3.10) $\lambda = z - 1/z + 1$ $(z^5 = \eta)$.

Table 1. *Rotation groups corresponding to modular equations of order b*

b	M	G	Description
2	3	6	$D(3)$ = dihedral (triangle)
3	4	12	$A(4)$ = tetrahedral
4	6	24	$S(4)$ = octahedral
5	6	60	$A(5)$ = icosahedral

Table 2. *Modular equation $J_b(j(\tau), j(b\tau)) = 0$ as found by eliminating η and η_0 in the equations $j(\tau) = \psi_b(\eta)$, $\eta_0 = \phi_b(\eta)$, $j(b\tau) = \psi_b(\eta_0)$*

b	$\psi_b(\eta)$	$\eta_0 = \phi_b(\eta)$
2	$64(\eta + 3)^3/(\eta - 1)^2$	$(\eta_0 - 1)(\eta - 1) = 1$
3	$27\eta(\eta + 8)^3/(\eta - 1)^3$	$(\eta_0 - 1)(\eta - 1) = 1$
4	$16(\eta^2 + 14\eta + 1)^3/\eta(\eta - 1)^4$	$\eta_0 + \eta = 1$
5	$-(\eta^2 - 10\eta - 5)^3/\eta$	$\eta_0\eta = 125$

Note: For $b = 2$, this table corresponds to equations (11.3.2a, b).

The rational parameter λ is needed for applications to ring class field theory.

Next we interpret the transformation of the z-sphere in terms of the rotation of regular solids. We illustrate this for $b = 2$ and then refer the remaining case to Tables 2, 3, and 4. First of all, the function $\psi_2(\eta)$ of Table 2 is one of a pair (connected by an elementary rational identity in η),

$$(11.3.11) \quad \begin{aligned} j(\tau) &= 64(\eta + 3)^3/(\eta - 1)^2 \\ &= 1728 + 64(\eta - 9)^2\eta/(\eta - 1)^2. \end{aligned}$$

Actually, we must transform $\eta = z^2$ (by Table 2) to find

$$(11.3.12a) \quad j(\tau) = 64F^3/V^2$$
$$(11.3.12b) \quad j(\tau) = 1728 + 64E^2/V^2$$

where polynomials are now defined:

$$(11.3.13a) \quad V = z^2 - 1 \quad \text{(roots } 1, -1, \infty)$$
$$(11.3.13b) \quad E = z^3 - 9z \quad \text{(roots } 0, 3, -3)$$
$$(11.3.13c) \quad F = z^2 + 3 \quad \text{(roots } \sqrt{-3}, -\sqrt{-3})$$

Table 3. *Use of a conjugate of η to solve for $j(b\tau)$ from $j(\tau)$ by the replacement of the equation*
$\eta_0 = \phi_b(\eta)$ *by* $\eta_0 = \phi_b(\eta')$ *where* $\eta = P(z)$, $z' = L(z)$, $\eta' = P(z')$

b	$P(z)$ [or $P(\lambda)$] $= \eta$	$z' = L(z)$ [or $\lambda' = L(\lambda)$]
2	$z^2 = \eta$	$(z' - 1)(z - 1) = 4$
3	$z^3 = \eta$	$(z' - 1)(z - 1) = 3$
4	$z^4 = \eta$	$(z' + 1)(z + 1) = 2$
5	$z^5 - 1/z^5 = -11 + 125/\eta$	$(z' + \epsilon)(z + \epsilon) = 2 + \epsilon, \epsilon = (1 + \sqrt{5})/2$
5(bis)	$(E(\lambda) =)\lambda^5 - 5\lambda^4 + 15\lambda^3 - 25\lambda^2 + 25\lambda$ $= 125/\eta \ [\lambda = (z - 1)/z + 1)]$	$\lambda\lambda' = 5$

Note: For $b = 5$(bis) the equations are $\eta = P(\lambda)$, $\lambda' = L(\lambda)$, and $\eta' = P(\lambda')$. For $b = 2$ this table corresponds to equations (11.3.3b) and (11.3.4a, b).

Table 4. *Explicit form of the conjugate* $\eta_0 = F_b(\eta)$ *found from Table 3*

b	$F_b(\eta)$	$G_b(Z)$
2	$[(5\eta - 9) + (\eta + 3)\eta^{1/2}]/8(\eta - 1)$	$(z + 3)^2/8(z + 1)$
3	$[(5\eta - 8) + (\eta - 4)\eta^{1/3} + 6\eta^{2/3}]/9(\eta - 1)$	$(z + 2)^3/9(z^2 + z + 1)$
4	$8\eta^{1/4}(1 + \eta^{1/2})/(1 + \eta^{1/4})^4$	$8(1 + z^2)/(1 + z)^4$
5	$125/E(1/E^{-1}(125/\eta))$	$(1 + z - 1/z)^6/(11 + z^5 - 1/z^5)$

Note: For $b = 2, 3, 4$, an alternate form of the conjugation is $\eta = z^b$, $\eta_0 = z_0^b$, and $z_0^b = G_b(z)$. For $b = 5$, the corresponding form is $11 + z_0^5 - 1/z_0^5 = G_5(z)$, and E^{-1} denotes the inverse of E defined in Table 3. For $b = 2$, this table corresponds to equations (11.3.5c) and (11.3.6c).

Of course, it may seem strange that V is considered to be of degree 3 whereas F is considered to be of degree 2, but this is explained by the homogeneity of the identity in (11.3.12a, b), namely

(11.3.13d) $$F^3 = E^2 + 27V^2.$$

Thus, V, E, and F are polynomials on the z-sphere whose roots are the vertices, the midpoints of the edges, and the centers of the faces of a polyhedron. In this case, it is a degenerate one: a dihedral (or two-faced) triangle represented by the upper and lower half-planes (as front and rear faces) with vertices taken as the roots of V, midpoints as the roots of E, and centers as the roots of F (one in each face). It is easily verified that the following transformations:

(11.3.14a) $\qquad z \to z' = -z$

(11.3.14b) $\qquad z \to z' = (z + 3)/z(z - 1)$

generate the group G of the dihedral triangle.

In Tables 5 and 6 some generating transformations are listed. The polyhedron in each case has v vertices (roots of V, one of which is always ∞), e edges (centered at roots of E), and f faces (centered at roots of F, the concept of "center" being noneuclidean). The transformations on z include trivial ones like (11.3.14a) (rotations of z that do not affect η) and nontrivial ones like (11.3.14b) [which generate the irreversible changes of root, as in equation (11.3.14b)]. The nontrivial ones are designated by (*).

11.3.15. Theorem (Klein). *A rotational function of η that is invariant under the rotation group G is the ratio of two polynomials of equal*

Table 5. *Transformations of the z-sphere that generate the rotation group G, showing actions on the roots of polynomials V, E, and F for vertices, edges, and faces (respectively) of the corresponding polyhedron*

b	Polynomials	Invariants	Generators of G
2	$V = (z^2 - 1)$ $E = (z^3 - 9z)$ $F = (z^3 + 3)$	$j(\tau) = 64F^3/V^2$ $j(\tau) - 1728 = 64E^2/V^2$ $(F^3 = E^2 + 27V^2)$	$z' = -z$ $z' = (z + 3)/(z - 1)$ (*)
3	$V = (z^3 - 1)$ $E = z^6 - 20z^3 - 8$ $F = z(z^3 + 8)$	$j(\tau) = 27F^3/V^3$ $j(\tau) - 1728 = 27E^2/V^3$ $(F^3 = E^2 + 64V^3)$	$z' = \omega z$ $z' = (z + 2)/(z - 1)$ (*)
4	$V = z(z^4 + 1)$ $E = (z^8 - 34z^4 + 1)(z^4 + 1)$ $F = z^8 + 14z^4 + 1$	$j(\tau) = 16F^3/V^4$ $j(\tau) - 1728 = 16E^2/V^4$ $(F^3 = E^2 + 108V^4)$	$z' = iz$ $z' = 1/z$ $z' = (1 - z)/(1 + z)$ (*)
5	$V = z(z^{10} + 11z^5 - 1)$ $E = z^{30} + 522z^{25} - 10{,}005z^{20}$ $\quad - 10{,}005z^{10} - 522z^5 + 1$ $F = z^{20} - 228z^{15} + 494z^{10}$ $\quad + 228z^5 + 1$	$j(\tau) = -F^3/V^5$ $j(\tau) - 1728 = -E^2/V^5$ $(F^3 = E^2 - 1728V^5)$	$z' = \omega z$ $z' = -1/z$ $z' = (-\epsilon z + 1)/(z + \epsilon)$ (*) $[\epsilon = \omega + 1/\omega = (1 + \sqrt{5})/2]$

Note: The generators that lead to conjugates of η in Tables 3 and 4 are denoted by (*). (Here $\omega = \exp 2\pi i/b$.) For $b = 2$, this table corresponds to equations (11.3.12a, b), (11.3.13d), and (11.3.14a, b).

Table 6. *Action of the generators of G (from Table 5) on the roots of V(z)*

b	Roots of $V(z)$	Generators of G	Action on roots
2	$A = -1, B = 1, C = \infty$	$z' = -z$ $z' = (z + 3)/(z - 1)$	$\left.\begin{array}{l}(AB) \\ (BC)\end{array}\right\} \Rightarrow D(3) = S(3)$
3	$A = 1, B = \omega, C = \omega^2,$ $D = \infty$	$z' = \omega z$ $z' = (z + 2)/(z - 1)$	$\left.\begin{array}{l}(ABC) \\ (AD)(BC)\end{array}\right\} \Rightarrow A(4)$
4	$A = 0, B = 1, C = -i,$ $D = \infty, E = -1, F = i$	$z' = iz$ $z' = 1/z$ $z' = (1 - z)/(1 + z)$	$(BFEC)$ $(AD)(CF)$ $(AB)(CF)(DE)$

[The permutations that constitute $S(4)$ are those of the four pairs of opposite triangles of the octahedron: $\{ABC, DEF\}, \{ABF, DEC\}, \{AEC, DBF\},$ and $\{AEF, DBC\}$.]

5	$A = \{0, \infty\}, B = \{-\epsilon, -\epsilon'\},$ $C = \{-\epsilon\omega, -\epsilon'\omega^4\}, D = \{-\epsilon\omega^2, -\epsilon'\omega^4\},$ $E = \{-\epsilon\omega^3, -\epsilon'\omega^2\}, F = \{-\epsilon\omega^4, -\epsilon'\omega\}$	$z' = \omega z$ $z' = -i/z$ $z' = (-\epsilon z + 1)/(z + \epsilon)$	$(BCDEF)$ Identity $(AB)(CF)(DE)$

[The permutations that generate the icosahedral group, $A(5)$, are shown acting on the six axes A, B, C, D, E, and F.]

Note: The parameters are defined in Table 5, except for $\epsilon' = (1 - \sqrt{5})/2$. For $b = 2$, this table corresponds to equations (11.3.13a) and (11.3.14a, b).

weight in V, E, and F (of weight, υ, e, and f, respectively). Hence, such a function is a rational function of j(τ).

A more general result involves the use of homogeneous polynomials obtained by setting $z = z_1/z_2$. Thus,

$$(11.3.16a) \qquad V_0(z_1, z_2) = V(z_1/z_2)z_2^v$$

$$(11.3.16b) \qquad E_0(z_1, z_2) = E(z_1/z_2)z_2^e$$

$$(11.3.16c) \qquad F_0(z_1, z_2) = F(z_1/z_2)z_2^f.$$

Thus, the root ∞ of V is explicitly evident. [For $b = 2$, $V_0 = (z_1^2 - z_2^2)z_2$, which vanishes for $z_1/z_2 = \pm 1$ and also for $z_2/z_1 = 0$ or equivalently $z_1/z_2 = \infty$.] Then the transformations of G become linear homogeneous; that is, $z' = (z + 3)/(z - 1)$ becomes

$$(11.3.16d) \qquad z_1' = z_1 + 3z_2 \qquad z_2' = z_1 + z_2.$$

11.3.17. Theorem (Klein) *A homogeneous polynomial $P(z_1, z_2)$ is covariant over the group G (i.e., it becomes multiplied by constant under operations of G) if and only if P is an isobaric polynomial in V_0, E_0, and F_0.*

We conclude by sketching a derivation of equation (11.3.2a) for $b = 2$. We begin by considering the function

$$(11.3.18a) \qquad j(2\tau) = F(\tau).$$

It satisfies [by the invariance of $j(\tau)$]

$$(11.3.18b) \qquad F(\tau + 1) = F((\tau - 1)/(2\tau - 1)) = F(\tau)$$

[for example, $j(2(\tau - 1)/(2\tau - 1)) = j(1 - 1/(2\tau - 1)) = j(2\tau - 1)$, etc.]. These transformations of $F(\tau)$ belong to a fundamental domain $\mathcal{D}^3$ that consists of three replicas of $\mathcal{D}$, the fundamental domain of $j(\tau)$.

We verify this assertion by noting that the boundaries of $\mathcal{D}^3$ are identified under the operations of equation (11.3.18b), whereas no smaller fundamental domain can suffice because there are three conjugates of $j(2\tau)$ for each value of $j(\tau)$. Since $\mathcal{D}^3$ is of genus 0, there is a function μ mapping $\mathcal{D}^3$ onto the sphere so that $\mu(A) = -1$, $\mu(K) = 0$, and $\mu(\infty) = \infty$ (see Figure 11.1). Also, $j(\tau)$ (having three conjugates) must be a cubic polynomial in μ. If we count the number of regions that meet at

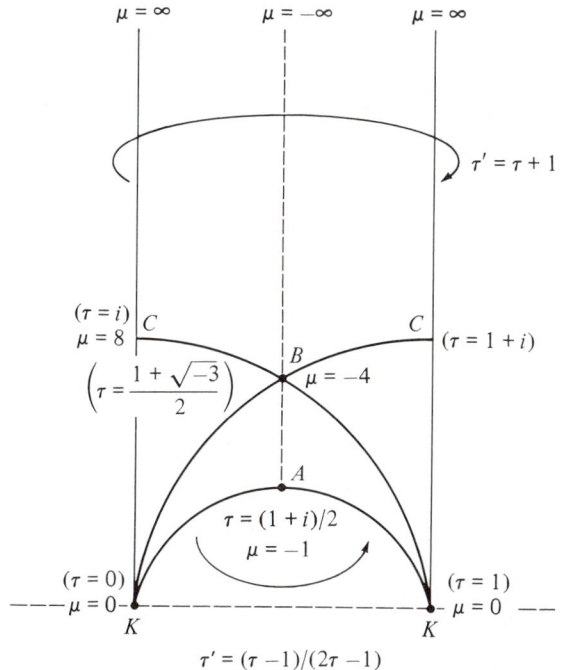

Figure 11.1. Fundamental domain $\mathcal{D}^3$ for $j(2\tau)$, showing how μ is single-valued whereas $j(\tau)$ is triple-valued.

various points, we see that B is a triple point, whereas K and C are double points. Thus,

(11.3.19a) $\qquad j(B) = 0 \leftrightarrow \mu = b, b, b$

(11.3.19b) $\qquad j(K) = \infty \leftrightarrow \mu = 0, 0, \infty$

(11.3.19c) $\qquad j(C) = 1728 \leftrightarrow \mu = -1, c, c$

for suitable constants b and c. By continuity, we see that

(11.3.19d) $\qquad\qquad b < 0 \quad\text{and}\quad c > 0.$

This leads to the form

(11.3.19e) $\qquad\qquad j(\tau) = (\mu - b)^3/\mu^2 \,(\text{const}).$

It is not hard to verify that the conditions on multiple roots at $\mu = c$ and $\mu = 0$ lead to $b = -4$ and $c = 8$. Thus,

(11.3.20a) $\qquad\qquad j(\tau) = 64(\mu + 4)^3/\mu^2 [= S_3(\mu)].$

But $j(2\tau)$ is rational in μ (the global parameter for $\mathcal{D}^3$) so

$$(11.3.20b) \qquad j(2\tau) = S_3(\mu')$$

where μ' is another global parameter for the same region. The reason for this curious symmetry is that $j(\tau/2)$ is a conjugate of $j(2\tau)$ in the transition from equation (11.3.20a) to (11.3.20b) by a different choice of root μ. Thus, this pair of equations could be replaced by

$$(11.3.21) \qquad j(\tau) = S_3(\mu'), \qquad j(\tau/2) = R_3(\mu').$$

Clearly, μ and μ' as two global parameters must be rationally related of degree 1 (linear fractional). This relation can be determined from the expansions at ∞ ($q \to \infty$), as follows:

$$(11.3.22a) \qquad j(\tau) = 1/q + 744 + \cdots = 64\mu + 12 \cdot 64 + \cdots$$
$$(\mu \to \infty)$$
$$(11.3.22b) \quad j(\tau) = 64 \cdot 64/\mu^2 + \cdots \qquad (\mu \to 0)$$

and $j(2\tau)$ from equation (11.3.20b) has a similar expansion in μ'. But

$$(11.3.23a) \qquad j(2\tau) = 1/q^2 + 744 + \cdots = 64\mu' + 12 \cdot 64 + \cdots$$
$$(\mu' \to \infty)$$
$$(11.3.23b) \quad j(2\tau) = 64 \cdot 64/\mu'^2 + \cdots \qquad (\mu' \to 0).$$

Obviously, equations (11.3.22a, b) and (11.3.23a, b) relate μ and μ'. It is left as an exercise to see that the only linear relation possible is $\mu\mu' = 1$. Finally we have the modular equation of Klein [note in equation (11.3.2a) that $\eta - 1$ is μ]:

$$(11.3.24a) \qquad \qquad j(\tau) = 64(\mu + 3)^3/\mu^2$$
$$(11.3.24b) \qquad \qquad j(2\tau) = 64(\mu' + 3)^3/\mu'^2$$
$$(11.3.24c) \qquad \qquad \mu\mu' = 1.$$

The results for $b = 3, 4, 5$ are similar but more intricate. We remark that the choice of η rather than μ was based on a notational convenience, for it is $\eta^{1/b}$ that enters directly into the iteration process. Actually μ is more significant in function theory because μ has no zeros or poles except where $j = \infty$. In general, such parameters have "easy" relations (such as $\mu\mu' = 1$). For $b = 2$ this fact enables us to write μ explicitly. We recall [from equations (10.4.6a, b)] that

$$(11.3.25a) \quad \Delta(2\tau)2^{12}/\Delta(\tau), \qquad \Delta(\tau/2)/\Delta(\tau), \qquad \Delta((\tau + 1)/2)/\Delta(\tau)$$

belong to the same respective Reimann surfaces [over $j(\tau)$] as

$$(11.3.25b) \qquad j(2\tau), \qquad j(\tau/2), \qquad j((\tau + 1)/2).$$

At the same time these functions in (11.3.25b) have no zeros or poles except when $j = 0$ or ∞. To identify parameters, note from equation (11.3.24a) that

(11.3.25c) $\qquad \mu = j/64 + \cdots = 1/(64q) + \cdots$

(11.3.25d) $\qquad \Delta(2\tau)2^{12}/\Delta(\tau) = 64^2q + \cdots.$

Therefore we identify

(11.3.26) $\qquad 1/\mu = 64\Delta(2\tau)/\Delta(\tau) = 64q \prod_{m=1}^{\infty} (1 + q^m)^{24}.$

This justifies the definition of $j(\tau)$ in Chapter 1:

(11.3.27) $\quad j(\tau) = (1 + 256q\, \Pi\, (1 + q^m)^{24})^3/q\, \Pi\, (1 + q^m)^{24}.$

The functions in (11.3.25a) are classically known as "Weber" functions and are written as f_2^{24}, f_1^{24}, and $-f^{24}$, respectively. Thus, we have the expansions

(11.3.28a) $\qquad f(\tau) = q^{-1/48}\, \Pi\, (1 + q^{m-1/2})$

(11.3.28b) $\qquad f_1(\tau) = q^{-1/48}\, \Pi\, (1 - q^{m-1/2})$

(11.3.28c) $\qquad f_2(\tau) = 2^{1/2}q^{1/24}\, \Pi\, (1 + q^m)$

(where m, as usual, goes from 1 to ∞). These functions are mercifully smaller in numerical value and even simpler in appearance than $j(\tau)$ (so Weber found them easier to tabulate). To retrieve $j(\tau)$, we use

(11.3.29) $\quad j^{1/3} = (f^{24} - 16)/f^8 = (f_1^{24} + 16)/f_1^8 = (f_2^{24} + 16)/f_2^8.$

11.3.30. Remark. In Chapter 1 and the exercises that follow we offer illustrations of how to test for prime values p that satisfy

$$p \ (\text{or } 4p) = x^2 + \Delta b^{2t}y^2, \qquad x, y, t, \Delta \in \mathbf{Z}, \qquad \Delta > 0, \qquad t \geq 0$$

for given Δ as t varies. The method is to test for the splitting of p in a succession of class fields such as $\mathbf{Q}(\sqrt{-\Delta}, j(b^t\sqrt{-\Delta}))$ constructed iteratively with t. Because the fields are so complicated (not to mention the integer rings of each field), we check the splitting of p by a succession of generating equations modulo p. In principle, this process could have a flaw, in that there are always exceptional p (say) divisors of root discriminants or values that make some denominator vanish. Yet these exceptional p not merely are finite in number but, in the cases covered here, they do not occur! If the iteration process can be *started* modulo p, then p is never an exceptional value when $b = 2, 3, 4, 5$. This fact is treated in detail in the references.

11.4. Epilogue to Weber's theorem

It is historically inevitable that a result as momentous as Weber's theorem must search for its natural role in a larger theory. The nature of this larger theory seems to be Hilbert's twelfth problem, the construction of algebraic number fields (as generally as possible) by the use of transcendental functions (in which some algebraic argument is substituted). This problem is still not very satisfactorily solved, but two aspects deserve some brief mention as appropriate to the construction of class fields, namely the Kronecker Jugendtraum and (even more briefly) the Hecke modular functions.

Kronecker conjectured (correctly) that all abelian fields $K/\mathbf{Q}$ are generated by roots of unity, that is, for some (conductor) m determined by K:

$$(11.4.1a) \qquad\qquad K \subseteq \mathbf{Q}(\exp 2\pi i/m).$$

Kronecker also conjectured (analogously but incorrectly) that all abelian extensions K/k, where $k = \mathbf{Q}(\sqrt{d_1})$ (with d_1 a negative field discriminant), are generated by abelian fields together with ring class fields. Thus, for some m, depending on K and k,

$$(11.4.1b) \qquad K \subseteq K^*(m) = k(j((d + \sqrt{d})/2), \exp 2\pi i/m)$$

for $d = m^2 d_1$. The first (correct) result, of course, appeared as an immediate consequence of the WT-correspondence (Theorem 7.4.18); the second (incorrect) result is known as Kronecker's "Jugendtraum." It is interesting to see how close to the truth it is.

11.4.2. Theorem (Hasse). *If K is relatively abelian over k with the conductor dividing the rational integer m, then*

$$(11.4.2a) \qquad k \subseteq K \subseteq K_m^l, \qquad K_m^l/k \leftarrow WT \rightarrow J_m^l$$
$$(11.4.2b) \qquad J_m^l = \langle a \equiv 1 \ mod^\times m \rangle$$

by the usual concepts of unit ray class field and unit ideal group. Then, referring to $K^*(m)$ in equation (11.4.1b),

$$(11.4.2c) \quad K^*(m)/k \leftarrow WT \rightarrow J^*(m)$$
$$= \{a \equiv r \ mod^\times m, \ r \in \mathbf{Z}, \ r^2 \equiv 1 \ mod \ m\}.$$

Accordingly, K_m^l is approximated by $K^*(m)$ to within a composite quadratic field:

$$(11.4.2d) \qquad (K \subseteq) K_m^l = K^*(m)(\sqrt{\mu_1}, \ldots, \sqrt{\mu_t})$$

for $\mu_i \in K^*(m)$. (In fact, $1 + t$ is the number of prime factors in m if m is odd or in $m/2$ if m is even.)

Proof. Note that the unit ring class field $K(m)$ satisfies

(11.4.2e) $K(m)/k \leftarrow \text{WT} \rightarrow \langle \mathfrak{a} \equiv r \bmod^\times m, r \in \mathbf{Z}, (r, m) = 1 \rangle.$

The abelian field $k(\exp 2\pi i/m)/k$ imposes the further splitting condition that p (in k) splits in $k (\exp 2\pi i/m)$ exactly when $N\mathfrak{p} \equiv 1 \bmod^\times m$ (see Illustration 7.5.8). This condition translates into the group condition $\mathfrak{a} \equiv r \bmod^\times m$, where $N(\mathfrak{a}) \equiv r^2 \bmod^\times m$. So $r^2 \equiv 1 \bmod m$ and the result (11.4.2c) follows [since $J^*(m)/J_m^1 = C(2)^t$]. $\qquad\square$

Weber, Fueter, Hasse, and many others have devised methods of directly generating the relatively abelian extension K/k. These methods are more difficult computationally and involve redefining $p(u)$, the elliptic function of Weierstrass (see Chapter 5) to make it homogeneous in its variable and period structure (the so-called Weber function). Using Hasse's notation, we start with $p(u)$ defined for the integer lattice of the field k and set

(11.4.3) $T(u) = \begin{cases} p^2(u)/g_2, & k = \mathbf{Q}(i) \\ p^3(u)/g_3, & k = \mathbf{Q}(\sqrt{-3}) \\ p(u)g_2^2g_3/(g_2^3 - 27g_3^2), & \text{other } k. \end{cases}$

Now the division points of the periods Ω, that is, Ω/M ($M \in \mathbf{Z}$), provide arguments for $T(u)$ that produce the algebraic numbers required to enlarge the ring class fields to ray class fields (K_m^1). The most significant result here is perhaps not the computation of class fields but the use of division points on the cubic curve to study the class structure of a quadratic field.

The first effort at imbedding Weber's theorem in a larger result was made by Hecke, who replaced $\mathbf{Q}$ by K_0, a totally real field, and replaced k by a totally imaginary field over K_0. If K_0 is of degree r, then the modular function field of Weber is replaced by a field of the degree of transcendence r of (Hecke) modular functions of r complex variables. It became clear very quickly that Weber's theorem has no direct generalization. When quadratic surds are substituted into these modular functions, the singular moduli obtained by analogy seem to consistently generate *proper* subfields of the ring class fields. Although the Hecke modular functions are now a thriving enterprise, the computational elegance a la Weber seems to be no longer continuing in center stage.

Exercises

11.1. Show that the exercises in Chapter 1 follow from the entries in Table 4 for discriminants -16.4^t and -8.4^t ($t = 0, 1, 2$).

11.2. Let $p \equiv 1 \bmod 3$, so $4p = x^2 + 27y^2$. Show that the t for which $3^t \parallel y$ is found from the inductive process where $a_1 = 8/9$, $r_s^3 \equiv a_s \bmod p$, and $a_{s+1} \equiv (r_s + 2)^3/9(r_s^2 + r_s + 1) \bmod p$. Then t is the last s for which r_s is definable mod p.

11.3. Let $p \equiv 1$ or $9 \bmod 20$, so $p = x^2 + 5y^2$. Show that the t for which $5^t \parallel y$ is found from the inductive process where $a_1^2 \equiv 125 \bmod p$ and $E(r_s) \equiv 125/a_s$, $a_{s+1} \equiv E(5/r_s) \bmod p$. Again t is the last s for which r_s is definable mod p. [Here $E(\lambda)$ is the polynomial in Table 3.]

11.4. In Table 4, verify the group G by its description as $D(3)$, $A(4)$, $S(4)$, and $A(5)$, respectively.

11.5. Verify (11.3.20a) from equations (11.3.19a–d).

11.6. Find $j(2i)$, $j(4i)$, $j(\sqrt{-3})$, and $j(2\sqrt{-3})$. *Hint:* First solve for η.

11.7. Verify that the normal closure of $\mathbf{C}(\eta)$ over $\mathbf{C}(j)$ has the Galois group $\Gamma(b)$ for $b = 2, 3, 4, 5$ from Exercise 10.13.

11.8. Verify the following values of Weber's functions, $f(i) = 2^{1/4}$, $f_1(\sqrt{-2}) = 2^{1/4}$, $f(\sqrt{-3}) = 2^{1/3}$, by using the values of j.

11.9. Show that for $k = \mathbf{Q}(i)$, the ray class field

$$K_8^1 \leftarrow\text{WT}\rightarrow J_8^1 = \langle \mathfrak{a} \equiv 1 \bmod^\times 8 \rangle$$

is given by $K_8^1 = K(8) \times \mathbf{Q}(\exp 2\pi i/16) = \mathbf{Q}(i, j(8i), \exp 2\pi i/16)$. Verify that $K(8) = k(2^{1/4})$, $\mathbf{Q}(\exp 2\pi i/16) = k(\sqrt{2 + \sqrt{2}})$,

$$\text{Spl}(K(8)) = \{p = x^2 + 64y^2\}$$
$$\text{Spl}(\mathbf{Q}(\exp 2\pi i/16)) = \{p \equiv 1 \bmod 16\}$$
$$\text{Spl}(K_8^1) = \{p = x^2 + 64y^2; x \equiv +1 \bmod 8\}.$$

Also Gal $K_8^1/\mathbf{Q} = G = \langle S, T, W : S^4 = T^2 = W^2 = [W, S] = [W, T] = 1, [S, T] = W \rangle$ (of order 16). Here T is a complex conjugation and $Si = i$, $S2^{1/4} = i2^{1/4}$, and $S\sqrt{1 \pm \sqrt{2}} = \sqrt{1 \mp \sqrt{2}}$. Also $G' = \langle W \rangle \leftarrow$ Gal$\rightarrow \mathbf{Q}(\exp 2\pi i/16)$, the maximal abelian subfield, so $\langle S^2, W \rangle \leftarrow$Gal$\rightarrow \mathbf{Q}(i, \sqrt{2})$ and $\langle S, W \rangle \leftarrow$Gal$\rightarrow k$; that is, Gal $K/k = C(4) \times C(2) = J_8^*/J_8^1$.

Bibliographic note

The general references of Chapter 10 still apply, plus (for complex multiplication) Cassels (1966), Cohn (1979a, 1982), and Iwasawa (1966). Special modular invariants are computed by Berwick (1928) and Herz (1966). For Remark 11.2.12, see Fricke (1928, p. 331). The icosahedral connection of Klein (1884) is developed in Cohn (1981b) parallel to results in Fricke (1928), Fricke and Klein (1890), and Stark (1973). For Remark 11.3.30, see Cohn 1981b. References relative to the Jugendtraum are Hasse (1926) and Langlands (1976); references relative to higher fields and modular functions of several variables are Cohn (1983), Hecke (1912), Shimura (1968), and Shimura and Taniyama (1961).

References

The standard historical bibliographies are to be found in Hilbert (1897), Hasse (1930b), and Narkiewicz (1974) (which also serves as a source reference). A computer-oriented bibliography appears in Zimmer (1972).

Ahlfors, L. V., and Sario L. 1960. *Riemann Surfaces.* Princeton: Princeton University Press.

Artin, E. 1924. Über eine neue Art der L-Reihen. *Abh. Math. Sem. Univ. Hamburg* 3:89–108.

Artin, E. 1927. Beweis des allgemeinen Reciprozitätsgesetzes. *Abh. Math. Sem. Univ. Hamburg* 5:353–63.

Artin, E. 1932. Lecture notes on class field theory (ed. O. Taussky, Appendix to Cohn 1978). Original lecture notes, Göttingen, 1932.

Artin, E., and Tate, J. 1952. *Algebraic Numbers and Algebraic Functions.* New York: Gordon and Breach. Original lecture notes, New York University.

Berwick, W. E. H. 1928. Modular invariants expressible in terms of quadratic and cubic irrationalities. *Proc. Lond. Math. Soc.* 28:53–69.

Birch, B. J. 1973. Some calculations of modular relations (in Kuyk 1973).

Borel, A., Chowla, S., Herz, C. S., Iwasawa, K., and Serre, J.-P. (eds.). 1966. *Seminar on Complex Multiplication, Lecture Notes 21.* Berlin: Springer.

Browder, F. (ed.). 1976. Mathematical developments arising from Hilbert problems. *Proc. Symp in Pure Math.* 28 (2 vols.). Providence: American Mathematics Society.

Bruckner, G. 1966. Charakterisierung der galoisschen Zahlkörper deren zerlegte Primzahlen durch binäre quadratischen Formen gegeben sind. *Mach. Nach.* 32:317–326.

Cassels, J. W. S. 1966. Diophantine equations with special reference to elliptic curves. *J. Lond. Math. Soc.* 41:193–291.

Cassels, J. W. S., and Fröhlich, A. (eds.). 1967. *Algebraic Number Theory (Symposium).* Washington, D.C.: Thompson Book Co.

Chebotarev, N. G. 1926. Die Bestimmung der Dichtigkeit einer Menge von Primzahlen welcher zu einer gegebener Substitutionsklasse gehören. *Math. Ann.* 95:191–228.

Chebotarev, N. G. 1930. Zur Gruppentheorie des Klassenkörpers. *J. Reine Angew. Math.* 161:179–193.

Cohn, H. 1978. *A Classical Invitation to Algebraic Numbers and Class Fields*, with appendices by Olga Taussky. New York: Springer.

Cohn, H. 1979a. *Diophantine Equations over C(t) and Complex Multiplication*, number theory Carbondale (1979) (ed. M. Nathanson), lecture notes 751. New York: Springer.

Cohn, H. 1979b. Quaternionic compositum-genus. *J. Number Theory* 11:399–411.

Cohn, H. 1979c. Cyclic-sixteen class fields for $Q(\sqrt{-p})$ by modular arithmetic. *Math. Comp.* 33:1307–16.

Cohn, H. 1981a. The explicit Hilbert 2-cyclic class field for $Q(\sqrt{-p})$. *J. Reine Angew. Math.* 321:64–77.

Cohn, H. 1981b. Iterated ring class fields and the icosahedron. *Math. Ann.* 255:107–22.

Cohn, H. 1982. Introductory remarks on complex multiplication. *Int. J. Math. Math. Sci.* 5:675–90.

Cohn, H. 1983. Some examples of Weber–Hecke ring class field theory. *Math. Ann.* 265:83–100.

Cohn, H., and Barrucand, P. 1969. Note on primes of the form $x^2 + 32y^2$, class number and residuacity. *J. Reine Angew. Math.* 238:67–70.

Cohn, H., and Cooke, G. 1976. Parametric form of an eight class field. *Acta Arith.* 30:367–77.

Cohn, H., and Lagarias, J. 1983. On the existence of fields governing the 2-invariants of the class group of $Q(\sqrt{dp})$ as p varies. *Math. Comp.* 41:711–30.

Dedekind, R. 1871. Über die Theorie des ganzen algebraischen Zahlen, Suppl. XI to Dirichlet's *Vorlesungen über Zahlentheorie* (1894). Braunschweig: Vieweg.

Deuring, M. 1958. Die Klassenkörper der komplexen Multiplikation. *Enzykl. der Math. Wiss.*, Bd. 1.2.

Dickson, L. E. 1939. *Modern Elementary Theory of Numbers*. Chicago: University of Chicago Press.

Dirichlet, P. G. L. 1894. *Vorlesungen über Zahlentheorie* (appendices by Dedekind). Braunschweig: Vieweg.

Edwards, H. M. 1977. *Fermat's Last Theorem*. New York: Springer.

Fricke, R. 1928. *Lehrbuch der Algebra, III Algebraischen Zahlen*. Braunschweig: Teubner.

Fricke, R., and Klein, F. 1890. *Vorlesungen über die Theorie der elliptischen Modulfunktionen*. Leipzig: Teubner.

Fröhlich, A. 1959. The genus field and genus groups in finite number fields. *Mathematika* 6:40–6.

Fröhlich, A. (ed.). 1977. *Algebraic Number Fields, L-functions, and Galois Properties. (Proc. Durham Symp. 1975)*. London: Academic Press.

Fröhlich, A. 1983. *Central Extensions, Galois Groups, and Ideal Class Groups of Number Fields*, Contemporary Mathematics Series, vol. 24. Providence: American Mathematics Society.

Fueter, R. 1924. *Vorlesungen über die singulären Moduln und die komplexe Multiplikation der elliptischen Funktionen*, 2 vols. Berlin: Teubner.

Furtwängler, P. 1916. Uber das Verhalten der Ideale des Grundkörpers in Klassenkörper. *Monatsh. Math. Phys.* 27:1–15.

Gundlach, K. B. 1972. *Einführung in die Zahlentheorie*. Mannheim: Bibliographisches Institut.

Hall, M. Jr. 1959. *The Theory of Groups*. New York: Macmillan.

Hasse, H. 1926. Bericht uber neuere Untersuchungen und Probleme aus der algebraische Zahlkörper I. *Jahresber. Deutsch. Math. Verein* 35:1–55.

Hasse, H. 1927a. Bericht über neuere Untersuchungen und Probleme aus der algebraische Zahlkörper Ia. *Jahresber. Deutsch. Math. Verein* 36:233–311.

Hasse, H. 1927b. Neue Begründung der komplexen Multiplikation. *J. Reine Angew. Math.* 157:115–39.

Hasse, H. 1930b. Bericht über neuere Untersuchungen und Probleme aus der algebraische Zahlkörper II. *Jahresber. Deutsch. Math. Verein,* Erg. Bd. 6:1–201.

Hasse, H. 1930c. Führer, Diskriminante und Verzweigungskörper relativ-abelscher Zahlkörper. *J. Reine Angew. Math.* 162:169–184.

Hasse, H. 1930d. Arithmetische Theorie der kubischen Zahlkörper auf Klassenkörpertheoretischer Grundlage. *Math. Zeit* 31:565–82.

Hasse, H. 1931. Neue Begründung der komplexen Multiplikation. *J. Reine Angew. Math.* 165:64–88.

Hasse, H. 1933. Klassenkörpertheorie. Original lecture notes, Marburg.

Hasse, H. 1949. Invariante Kennzeichnung relativ-abelscher Zahlkörper mit vorgegbener Galoisgruppe uber einem Teilkörper des Grundkörpers. *Abh. Deutsch. Akad. Wiss. zu Berlin,* (Jahrgang 1947).

Hasse, H. 1967. History of class field theory (in Cassels and Fröhlich 1967).

Hecke, E. 1912. Zur Theorie der Modulfunktionen von zwei Variabeln und ihre Anwendung auf die Zahlentheorie. *Math. Ann.* 71:1–37.

Hecke, E. 1917. Über eine neue Anwendung der Zetafunktion auf die Arithmetik der Zahlkörper. *Nach. Ges. Wiss. Gottingen* 299–318.

Hecke, E. 1923. *Vorlesungen über die Theorie der algebraischen Zahlen.* Leipzig: Teubner.

Heilbronn, H. 1967. Zeta-functions and L-functions (in Cassels and Fröhlich 1967).

Herbrand, J. 1936. Le développement moderne de la théorie des corps algébriques. *Mem. Sci. Math.* 75. Paris: Gauthier-Villars.

Herz, C. S. 1966. Construction of class fields (in Borel et al. 1966).

Hilbert, D. 1897. Die Theorie der algebraischen Zahlkörper. *Jahrsber. Deutsch. Math. Ver.* 4:175–546.

Hilbert, D. 1899. Über die Theorie des relativquadratischen Zahlkörpers. *Math. Ann.* 51:1–127.

Holzer, L. 1966. *Klassenkörpertheorie.* Leipzig: Teubner.

Ishida, M. 1976. *The Genus Fields of Algebraic Number Fields, Lecture Notes 555.* Berlin: Springer.

Iwasawa, K. 1966. Class fields (in Borel et al. 1966).

Jacobson, N. 1974. *Basic Algebra I.* San Francisco: Freeman.

Jones, B. W. 1950. *The Arithmetic Theory of Quadratic Forms.* New York: Wiley.

Kisilevsky, H. 1970. Some results on Hilbert's Theorem 94. *J. Number Theory* 2:199–206.

Kisilevsky, H. 1976. Number fields with class number congruent to 4 mod 8, and Hilbert's Theorem 94. *J. Number Theory* 8:271–9.

Klein, F. 1884. *Vorlesungen über das Icosaeder und die Auflösung der Gleichung vom fünften Grade.* Leipzig: Teubner.

Kuyk, W. (ed.). 1973. *Modular Functions of One Variable I,* Proceedings International Summer School, University of Antwerp 1972, Lecture Notes 320. Berlin: Springer.

Lagarias, J. C. 1983. Sets of primes determined by systems of polynomial congruences. *Ill. J. Math.* 27:224–39.

Lagarias, J. C., and Odlyzko, A. M. 1977. Effective versions of the Chebotarev density theorem (in Fröhlich 1977).

Landau, E. 1918. *Einführung in die elementare und analytische Theorie der Algebraischen Zahlen und die Ideale.* Leipzig: Teubner.

Lang, S. 1970. *Algebraic Number Theory*. Reading, Mass.: Addison Wesley.

Langlands, R. P. 1976. Some contemporary problems with origins in the Jugendtraum (Hilbert's Problem 12) (in Browder 1976).

Lenstra, H. W. 1975. Euclid's algorithm in cyclic fields. *Lond. Math. Soc.* (2)10:457–65.

Masley, J. 1977. Odlyzko bounds and class number problems (in Fröhlich 1977).

Meyer, C. 1957. *Die Berechnung der Klassenzahl abelscher Körper über quadratischen Zahlkörpern*. Berlin: Akademie.

Minkowski, H. 1896. *Geometrie der Zahlen*. Leipzig: Teubner.

Morton, P. 1982. Density results for the 2-class groups of imaginary quadratic fields. *J. Reine Angew. Math.* 332:156–87.

Narkiewicz, W. 1974. *Elementary and Analytic Theory of Algebraic Numbers*. Warsaw: Polish Scientific Publishers.

Noether, E. 1921. Idealtheorie in Ringbereichen. *Math. Ann.* 83:24–66.

Odlyzko, A. M. 1977. On conductors and discriminants (in Frölich 1977).

Ogg, A. 1973. Survey of modular functions of one variable (in Kuyk 1973).

Rankin, R. A. 1969. *The Modular Group and Its Subgroups, Lecture Notes*. Madras: Ramanujan Institute.

Rédei, L., and Reichardt, H. 1934. Die Anzahl der durch 4 teilbaren Invarianten der Klassengruppe eine beliebigen quadratischen Zahlkörpers. *J. Reine Angew. Math.* 170:69–74.

Riebenboim, P. 1972. *Algebraic Numbers*. New York: Wiley-Interscience.

Satge, P. 1977. Décomposition des nombres premiers dans les extensions non abéliennes. *Ann. Inst. Fourier, Grenoble* 27:1–8.

Scholz, A. 1934. Über die Lösbarkeit der Gleichung $t^2 - Du^2 = -4$. *Math. Zeit.* 39:1935.

Serre, J.-P. 1966. Modular forms (in Borel et al. 1966).

Shanks, D. 1971. Class number, a theory of factorization, and genera, *Proc. Symp. Pure Math.*, vol. 20. Providence: American Mathematics Society.

Shimura, G. 1968. *Automorphic Functions and Number Theory, Lecture Notes 54*. Berlin: Springer.

Shimura, G., and Taniyama, Y. 1961. *Complex Multiplication of Abelian Varieties and Its Application to Number Theory*. Tokyo: Mathematics Society of Japan.

Stark, H. M. 1973. Class numbers of complex quadratic fields (in Kuyk 1973).

Swinnerton-Dyer, H. P. F. 1974. *Analytic Theory of Abelian Varieties, Lecture Notes 14, London Mathematics Society*. Cambridge: Cambridge University Press.

Takagi, T. 1920. Über eine Theorie des relativ-Abelschen Zahlkörpers. *J. Coll. Sci. Univ. Tokyo* 41:1–133.

Tate, J. T. 1950. Fourier analysis in number fields and Hecke's zeta-function ("Thesis") (in Cassels and Fröhlich 1967).

Taussky, O. 1937. A remark on the class field tower. *J. Lond. Math. Soc.* 12:82–85.

Taussky, O. 1971. *Hilbert's Theorem 94, Computers in Number Theory*. London: Academic Press.

Van der Waerden, B. L. 1950. *Moderne Algebra*. Berlin: Springer.

Weber, H. 1886. Theorie der abelschen Zahlkörper. *Acta Math.* 8:193–263.

Weber, H. 1891. *Elliptische Funktionen und algebraischen Zahlen*. Braunschweig: Vieweg.

Weber, H. 1899. *Algebra II*. Braunschweig: Vieweg.

Weil, A. 1967. *Basic Number Theory*. New York: Springer.

Weil, A. 1983. *Number Theory. An Approach through History from Hammurapi to Legendre*. Boston: Birkhauser.

Witt, E. 1936. Konstruktion von galoischen Körpern der Charakteristik p zu vorgegebener Gruppen der Ordnung p^f. *J. Reine Angew. Math.* 174:237–45.

Weyl, H. 1940. *Algebraic Theory of Numbers.* Princeton: Princeton University Press.

Whittaker, E. T., and Watson, G. N. 1940. *Modern Analysis.* Cambridge: Cambridge University Press.

Yui, N. 1978. Explicit form of the modular equation. *J. Reine Angew. Math.* 299/300:185–200.

Zimmer, H. G. 1972. *Computational Problems, Methods, and Results in Algebraic Number Theory, Lecture Notes 262.* Berlin: Springer.

Index